The
Introductory Differential Equations Companion

The Introductory Differential Equations Companion

by Scott Surgent

Copyright © 2022 Scott Surgent. Printed and bound in the United States of America. All Rights Reserved. No part of this book may be reproduced or transmitted in any form or by any means, electronic or mechanical, including photocopying, recording, or by an information storage or retrieval system without permission in writing from the publisher. Short passages used for the purposes of reviews are permitted.

ISBN-13: 9798849384115

www.surgent.net/debook

This is an original work, researched and written by the author.

Direction Field Plots and Phase Plane Plots contained within this book were created using software developed by, and with the permission and courtesy of, Dr. Ariel Barton of the University of Arkansas, Department of Mathematical Sciences.

Cover art by the author

All other graphs were rendered by the author.

Introduction

This book is designed to be an addendum for students taking an Introductory Differential Equations course at the collegiate level. It contains over a hundred detailed examples to help walk students through the concepts and some of the more challenging algebra that is typical for this course.

The topics contained herein are typical for most Introductory Differential Equations courses. Not all courses are identical so some of the topics may not be relevant for a particular student's experience, whereas this book may lack a topic being taught in another class elsewhere. Nevertheless, these topics comprise the bulk of most such courses, which tend to be fairly consistent among all schools.

This book, by design, is not aligned with any particular textbook. This book, by itself, is not a textbook. The emphasis is on detailed examples and practice, less so on the theory. Most textbooks on this subject do a fine job of covering the important theorems, and a student should access the textbook for these proofs. In some cases, a short proof, or outline of a proof, is given within this book, if it helps move the conversation along.

Most textbooks are limited by space to how many examples can be included and may exclude or compress steps along the way to ensure the examples "fit" within the pages. It is not uncommon for a problem in Differential equations to require many dozens of steps. The intention of this book is to show those steps in more detail if the main textbook cannot due to space concerns.

The examples within this book are approached by a combination of adherence to proper mathematical language, rigor and logic, as well as heuristic discussions that may help a student understand a concept better.

Students taking this course are assumed to have completed the full single-variable calculus sequence, but not necessarily have taken or started a multi-variable calculus course. Only one topic, Exact Equations, requires some knowledge of multi-variable calculus and this topic is moved to the appendix for that reason. Topics from Linear Algebra are used in some cases to solve certain differential equations. It is not necessary for a student to have already completed a course in Linear Algebra, and whenever necessary, a small amount of review is provided to explain the Linear Algebra concepts.

Students should have a solid understanding of the main calculus concepts as well as strong procedural and computational skills. If the student is unable to properly find a derivative or an aintiderivative of a function consistently, then that student will struggle mightily in this course.

The student will also discover that the calculus steps comprise just a small part of the computational process of a typical differential equation. The vast bulk of the work in solving a differential equation is algebra. At this level, poor algebra skills simply cannot be tolerated. The algebra can be daunting, but when carefully handled, is not difficult in the conceptual sense, just tedious at times. Thus, an incoming student to an Introductory Differential Equations course must have strong calculus skills and conceptual understanding, and flawless algebra skills. This is the challenge, and within this book, these steps are shown is as much detail as seems necessary to help students "get over the hump", so to speak.

I have taught this course many times over my 30 years of teaching at the collegiate level. The bulk of the contents within this book are derived from my own lecture notes and slides used in the classroom. I am aware of the challenges this course has on students and have used my experience to craft the examples so that the usual points of confusion are identified, explained and mitigated.

I don't advertise nor expect that this book will cure all ills and vault a student to the top of their class. It is my hope that these examples will help illuminate steps that may be causing a student confusion. Learning is incremental and often non-linear. The understanding often comes in bits and pieces. Thumbing through this book hopefully helps fill in holes of understanding and makes the student's experience a better one, perhaps the difference in a letter grade. First and foremost, I want students to succeed and learn a little more along the way.

If you see an error or have a suggestion, please visit my website for this book at

www.surgent.net/debook

There, you can look over any updates that may have already been suggested, while my email listed on the website allows you to contact me directly.

Remember, knowledge doesn't just fall out of the sky into your head. It takes concerted effort and practice over a long period to gain mastery. Take time to study the proofs and examples carefully. It is the only way.

Scott Surgent
September 2022

Table of Contents

1. Preliminary: A Review of Calculus 2
2. What is a Differential Equation and Verifying Solutions of Differential Equations 9
3. Classification of Differential Equations 13
4. Separation of Variables & Direction Fields . . . 16
5. First-Order Autonomous Differential Equations . . . 23
6. Integration Factors 31
7. Bernoulli Equations 36
8. Mixture Problems 40
9. Numerical Methods: Euler and Heun Methods . . . 47
10. Higher-Order Linear Homogeneous & Autonomous Differential Equations with Constant Coefficients . . . 53
11. Linear Independence of Solutions: The Wronskian . . 58
12. Complex Roots of the Auxiliary Polynomial 61
13. Repeated Roots of the Auxiliary Polynomial . . . 68
14. Reduction of Order 73
15. Undetermined Coefficients 79
16. Spring-Mass Systems 85
17. Cauchy-Euler Equations. 93
18. Laplace Transforms 97
19. Solving Initial Value Problems Using Laplace Transforms . 105
20. Laplace Transforms of Discontinuous Forcing Functions . 109
21. Using Laplace Transforms to Solve IVPs with Piecewise Forcing Functions 117
22. Impulse Forcing Function 123
23. Laplace Transforms of Special Cases 126
24. Laplace Transforms of Periodic Functions 132
25. Matrix Review: Determinants, Eigenvalues and Eigenvectors 136
26. Systems: Real Eigenvalues of Multiplicity 1 146
27. Systems: Complex Eigenvalues 152
28. Systems: Real Eigenvalues of Repeated Multiplicity . . 160

Free Extra Stuff!

The Imaginary Unit *i*, The Complex Plane and Roots of Unity . 164
Partial Fraction Decomposition 167
Variation of Parameters 177
Exact Equations 181
Series Solutions 185
Practice Problems with Solutions 196

Section 1
Preliminary: A Review of Calculus

To be successful in an introductory ordinary differential equations course, it is imperative to review the basic derivative and antiderivative forms from calculus. Most differential equations in an introductory course use calculus in a few steps, but the solution steps require a lot of algebra. In fact, it may be the algebra that is the most challenging aspect of an introductory ordinary differential equations course. Nevertheless, a strong aptitude with basic calculus is critical.

Derivatives

If $y = f(x)$ is a function, then its derivative is written as

$$y' = \frac{dy}{dx} = \frac{d}{dx}(f(x)) = f'(x).$$

All notation forms are equivalent. The form $\frac{d}{dx}$ is called an operator. By itself, it means nothing. But attached to a function, it means "find this function's derivative".

Constant Rule: $\frac{d}{dx} k = 0.$

Linear Rule: $\frac{d}{dx}(mx) = m.$

Power Rule: $\frac{d}{dx}(x^n) = nx^{n-1}$, where n is any real number.

Constant Multiplier Rule: $\frac{d}{dx}(kf(x)) = k \frac{d}{dx}(f(x)).$

Sum/Difference Rule: $\frac{d}{dx}(f(x) \pm g(x)) = \frac{d}{dx}(f(x)) \pm \frac{d}{dx}(g(x)).$

Linearity: $\frac{d}{dx}(k_1 f(x) \pm k_2 g(x)) = k_1 \frac{d}{dx}(f(x)) \pm k_2 \frac{d}{dx}(g(x)).$

Chain Rule: $\frac{d}{dx}(f(g(x))) = f'(g(x)) \cdot g'(x).$

Product Rule: $\frac{d}{dx}(f(x) \cdot g(x)) = f(x) \cdot \frac{d}{dx}(g(x)) + g(x) \cdot \frac{d}{dx}(f(x)).$

Quotient Rule: $\frac{d}{dx}\left(\frac{f(x)}{g(x)}\right) = \frac{g(x) \cdot f'(x) - f(x) \cdot g'(x)}{[g(x)]^2}.$

Exponential Rule (general): $\frac{d}{dx}(a^x) = a^x \ln a$, where $a > 0$ and $a \neq 1$.

Exponential rule (base-e): $\frac{d}{dx}(e^x) = e^x$, since $\ln e = 1$.

Logarithm Rule (general): $\frac{d}{dx}(\log_a x) = \frac{1}{x \cdot \ln a}$, where $a > 0$ and $a \neq 1$.

Logarithm Rule (base-e): $\frac{d}{dx}(\ln x) = \frac{1}{x}$, since $\ln x$ is log base-e, and $\ln e = 1$.

Sine and Cosine Rules: $\frac{d}{dx}(\sin x) = \cos x$ and $\frac{d}{dx}(\cos x) = -\sin x$.

Algebraic simplification should always be performed first to make the differentiation steps simpler and faster. Derivative forms are often written with a trailing dx to indicate the chain rule. For example, $\frac{d}{dx}(e^x) = e^x\,dx$, $\frac{d}{dx}(x^n) = nx^{n-1}\,dx$, $\frac{d}{dx}(\ln x) = \frac{1}{x}\,dx = \frac{dx}{x}$, and $\frac{d}{dx}(\sin x) = \cos x\,dx$.

Examples:

$\frac{d}{dx}(3x^2 + 2x - 5) = 6x + 2;$ $\qquad \frac{d}{dx}(\sqrt{x}) = \frac{d}{dx}(x^{1/2}) = \frac{1}{2}x^{-1/2} = \frac{1}{2\sqrt{x}}$

$\frac{d}{dx}\left(\frac{x^4 + 5x^3 - 6x^2 + x - 9}{x^2}\right) = \frac{d}{dx}\left(x^2 + 5x - 6 + \frac{1}{x} - \frac{9}{x^2}\right) = 2x + 5 - \frac{1}{x^2} + \frac{18}{x^3}$

$\frac{d}{dx}(x^3 + 4x + 7)^5 = 5(x^3 + 4x + 7)^4(3x^2 + 4)$

$\frac{d}{dx}(\sin 3x) = 3\cos 3x$

$\frac{d}{dx}(x^2 \cos 5x) = x^2(-5\sin 5x) + 2x(\cos 5x) = -5x^2 \sin 5x + 2x \cos 5x$

$\frac{d}{dx}(\tan x^2) = \frac{d}{dx}\left(\frac{\sin x^2}{\cos x^2}\right) = \frac{\cos x^2(2x\cos x^2) - \sin x^2(-2x\sin x^2)}{\cos^2 x^2}$

$\qquad = \frac{2x(\cos^2 x^2 + \sin^2 x^2)}{\cos^2 x^2} = \frac{2x(1)}{\cos^2 x^2} = 2x\sec^2 x^2$

$\frac{d}{dx}(e^{-7x^4}) = e^{-7x^4}(-28x^3) = -28x^3 e^{-7x^4}$

$\frac{d}{dx}(\ln(x^5 + 3x + 11)) = \frac{5x^4 + 3}{x^5 + 3x + 11}$

$\frac{d}{dx}((x^2 + 3)(2x - 5)) = \frac{d}{dx}(2x^3 - 5x^2 + 6x - 15) = 6x^2 - 10x + 6$

Antidifferentiation and Integration

An antiderivative of $y = f(x)$ is any function $F(x)$ such that $\frac{d}{dx}(F(x)) = f(x)$. It is differentiation performed in reverse. Antiderivatives are written using indefinite integral notation:

$$\int f(x)\,dx = F(x) + C,$$

where C is called the constant of integration. This means there is not just one antiderivative, but infinitely many, all "the same" except for a different trailing constant.

Constant Rule: $\int k\,dx = kx + C$

Power Rule: $\int x^n\,dx = \frac{1}{n+1} x^{n+1} + C$, for all n such that $n \neq -1$.

Natural logarithm Rule: $\int \frac{1}{x}\,dx = \ln x + C$, where $x > 0$. This takes care of the case from above when $n = -1$.

Constant Multiplier Rule: $\int kf(x)\,dx = k\int f(x)\,dx$.

Sum/Difference Rule: $\int (f(x) \pm g(x))\,dx = \int f(x)\,dx \pm \int g(x)\,dx$.

Linearity: $\int (k_1 f(x) \pm k_2 g(x))\,dx = k_1 \int f(x)\,dx \pm k_2 \int g(x)\,dx$

Substitution Rule: $\int f(g(x))g'(x)\,dx = f(g(x)) + C$. Often called *u-du* substitution.

Integration by Parts Rule: $\int u\,dv = uv - \int v\,du$.

Sine/Cosine Rules: $\int \cos x\,dx = \sin x + C$ and $\int \sin x\,dx = -\cos x + C$.

Definite Integral: $\int_a^b f(x)\,dx = F(b) - F(a)$.

Infinity as a bound: $\int_a^\infty f(x)\,dx = \lim_{b\to\infty} \int_a^b f(x)\,dx = \lim_{b\to\infty}(F(b) - F(a))$, assuming the limit exists.

Fundamental Theorem of Calculus (FTC)

If $F(x) = \int_a^x f(t)\, dt$, where f is continuous on some interval $a \leq x \leq b$, then $\frac{d}{dx} F(x) = f(x)$, where F is continuous on $a < x < b$.

$\int_a^b f(x)\, dx = F(b) - F(a)$, assuming $\frac{d}{dx} F(x) = f(x)$ over $a < x < b$.

If f is continuous on $a \leq x \leq b$, then the FTC guarantees the existence of a function F that satisfies the first part of the theorem. It does not show how such a function F can be found or expressed in terms of common functions (the so-called "closed form"). It is always permissible to express the function F as a function defined by the definite integral form. For example, $\int \sin(e^x)\, dx$ has an antiderivative but it is not expressible in closed form using common functions. Instead, its antiderivative is defined as $F(x) = \int_a^x \sin(e^t)\, dt$. Note that x is the independent variable of F, so that to avoid confusion (and "self-referencing"), a dummy variable t is used within the integrand.

Examples:

$\int x^2 + 5x - 3\, dx = \frac{1}{3}x^3 + \frac{5}{2}x^2 - 3x + C$

$\int \sqrt{x}\, dx = \int x^{1/2}\, dx = \frac{2}{3}x^{3/2} + C, \quad x \geq 0$

$\int (3x+5)(x^2 - 8)\, dx = \int (3x^3 + 5x^2 - 24x - 40)\, dx$

$\qquad\qquad\qquad\qquad\qquad = \frac{3}{4}x^4 + \frac{5}{3}x^3 - 12x^2 - 40x + C$

$\int \left(\frac{x^4 - 7x + 9}{x}\right) dx = \int \left(x^3 - 7 + \frac{9}{x}\right) dx = \frac{1}{4}x^4 - 7x + 9 \ln x + C, \quad x > 0$

$\int x(3x^2 + 7)^9\, dx \rightarrow \left.\begin{array}{l} u = 3x^2 + 7 \\ du = 6x\, dx \end{array}\right\} = \frac{1}{6} \int u^9\, du = \frac{1}{6}\left(\frac{1}{10} u^{10}\right)$

$\qquad\qquad\qquad\qquad\qquad\qquad\qquad = \frac{1}{60}(3x^2 + 7)^{10} + C$

$\int x^2 \cos x^3\, dx = \frac{1}{3} \sin x^3 + C$

$\int xe^{2x}\, dx \rightarrow \left.\begin{array}{ll} u = x & dv = e^{2x} \\ du = dx & v = \frac{1}{2}e^{2x} \end{array}\right\} = \frac{x}{2}e^{2x} - \frac{1}{4}e^{2x} + C$

$\int_{-1}^4 (x^2 - 1)\, dx = \left(\frac{1}{3}x^3 - x\right)\Big|_{-1}^4 = \left(\frac{1}{3}(4)^3 - (4)\right) - \left(\frac{1}{3}(-1)^3 - (-1)\right) = \frac{50}{3}$

Trigonometry

There are two common ways to define the cosine of x, abbreviated cos x, and the sine of x, abbreviated sin x. Below on the left is a circle centered at the origin, radius 1. This is called the **unit circle**. A ray is drawn from the origin outward. An angle t is defined to be the value swept from the positive x-axis to the ray, always in a counterclockwise direction (positive orientation). Since the circle has a perimeter of 2π, then the angle t is simply the fractional part of the perimeter. For example, the angle of 90°, which is swept out from the positive x-axis to the positive y-axis, is 1/4 of a circle. Thus, the angle t is 1/4 of the perimeter 2π, or $\pi/2$. This angle measurement is called **radian** measure. It is used exclusively in all analytical matters involving trigonometry.

The ray intersects the circle. The x-coordinate of this point is defined to be the cosine of t, or cos t. The y-coordinate is the sine of t, or sin t. The slope of the ray is the rise over run, and is defined to be the tangent of t, or tan t, where $\tan t = \dfrac{\sin t}{\cos t}$.

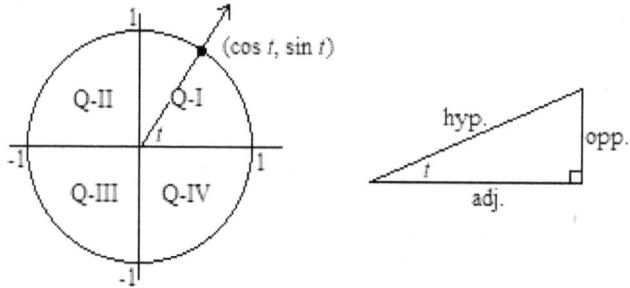

The cosine, sine and tangent values of angle t can also be found by a right triangle, where $\sin t = \dfrac{\text{opposite leg}}{\text{hypotenuse}}$, $\cos t = \dfrac{\text{adjacent leg}}{\text{hypotenuse}}$, and $\tan t = \dfrac{\text{opposite leg}}{\text{adjacent leg}}$. These two definitions are equivalent, as a right triangle can be drawn within the unit circle.

The circle is divided into 4 quadrants:

Quadrant	Radian Measure	Signs of cos t, sin t and tan t
Q-I:	$0 < t < \pi/2$	$\cos t > 0, \sin t > 0, \tan t > 0$
Q-II:	$\pi/2 < t < \pi$	$\cos t < 0, \sin t > 0, \tan t < 0$
Q-III:	$\pi < t < 3\pi/2$	$\cos t < 0, \sin t < 0, \tan t > 0$
Q-IV:	$3\pi/2 < t < 2\pi$	$\cos t > 0, \sin t < 0, \tan t < 0$

The common values for the cosine, sine and tangent of angles in the first quadrant are given below.

Angle degrees	Angle radians	$\cos t$	$\sin t$	$\tan t$
0°	0	1	0	0
30°	$\pi/6$	$\sqrt{3}/2$	$1/2$	$\sqrt{3}/3$
45°	$\pi/4$	$\sqrt{2}/2$	$\sqrt{2}/2$	1
60°	$\pi/3$	$1/2$	$\sqrt{3}/2$	$\sqrt{3}$
90°	$\pi/2$	0	1	undef.
180°	π	-1	0	0
270°	$3\pi/2$	0	-1	undef.

The cosine and sine functions are related by the Pythagorean Theorem. Two more corollary formulas can be defined.

$$\cos^2 t + \sin^2 t = 1; \qquad 1 + \tan^2 t = \sec^2 t; \qquad \cot^2 t + 1 = \csc^2 t.$$

The secant of t, written $\sec t$, is $\sec t = \frac{1}{\cos t}$; the cotangent of t, written $\cot t$, is $\cot t = \frac{1}{\tan t} = \frac{\cos t}{\sin t}$; and the cosecant of t, written $\csc t$, is $\csc t = \frac{1}{\sin t}$. The cotangent and cosecant functions are rarely used as they can always be written in terms of cosine and/or sine.

Common identities are the sum/difference formulas:

$$\sin(t \pm u) = \sin t \cos u \pm \cos t \sin u;$$
$$\cos(t \pm u) = \cos t \cos u \mp \sin t \sin u.$$

Note the reversal of the plus-minus sign in the cosine formula.

From these, the double angle formulas are developed:

$$\sin 2t = 2 \sin t \cos t; \qquad \cos 2t = \cos^2 t - \sin^2 t.$$

The half-angle formulas are

$$\cos^2 t = \frac{1}{2} + \frac{1}{2} \cos 2t; \qquad \sin^2 t = \frac{1}{2} - \frac{1}{2} \cos 2t.$$

Common characteristics of the cosine and sine functions are:

$y = \cos t$
    Domain: All real numbers, $-\infty < t < \infty$.
    Range: $-1 \leq y \leq 1$.
    Period: 2π, meaning that $\cos(t + 2\pi) = \cos t$ for all t.
    Symmetry: $\cos(-t) = \cos t$. Cosine is an even function.

$y = \sin t$
    Domain: All real numbers, $-\infty < t < \infty$.
    Range: $-1 \leq y \leq 1$.
    Period: 2π, meaning that $\sin(t + 2\pi) = \sin t$ for all t.
    Symmetry: $\sin(-t) = -\sin t$. Sine is an odd function.

$y = \tan t$
    Domain: All real numbers except $\pm k\pi/2$, where k is odd.
    Range: $-\infty < y < \infty$.
    Period: π, meaning that $\tan(t + \pi) = \tan t$ for all t in the domain.
    Symmetry: $\tan(-t) = -\tan t$. Tangent is an odd function.

Algebra Skills

The following is a listing of common algebra skills that need to be mastered and are relevant for a course like this:

Arithmetic: adding fractions, distributing, order of operations, reciprocation, odd and even radicals, and so on.
Rules of exponents.
Functions, notation and graphing.
Odd and even symmetry.
Domain and range of functions.
Polynomials: factoring and multiplying, evaluation thereof.
Natural exponential and logarithmic functions.
Simplifying expressions.
Solving for the unknown.
The imaginary root i and its various integers powers.
Matrices. Review is provided in this book as needed.
Partial fractions. Review will be provided as needed.

Many of the problems one will encounter in an introductory ordinary differential equations course involve a *lot* of algebra, and it is imperative that a student's algebra skills be sharp and accurate. One small error in an early step can cause the remaining steps to be false (*i.e.* wasted work). Go slow and carefully. Understand each step before moving onward.

Section 2
What is a Differential Equation and Verifying Solutions of Differential Equations

An **ordinary differential equation** is any equation that, given a function of a single independent variable $y = f(x)$, contains at least one of its derivatives (first, second, third and so on). The function itself may or may not be present in the equation. In many applications, time is often the independent variable, so the letter t is often used in this context.

The following are examples of differential equations of a single independent variable:

$$y' + 2y = 4$$

$$y'' + 3xy' + 4y = x^2$$

$$\frac{d^2y}{dx^2} + 6x\frac{dy}{dx} + y = 0$$

$$y' = 6x$$

$$3y'' + 6xy' = 4$$

Notation varies. For convenience, the equations are usually written using the short forms y, y', y'' and so on. Note also that the independent variable may not "appear" in the differential equation (such as in the first example above), but it is "present" as part of the eventual solution.

The function y itself may be absent as long as at least one of its derivatives is present. The fourth and fifth examples shown above lack y in its equation.

A **solution** of a differential equation is a set of functions that upon substitution, makes the equation true. A differential equation may have initial condition(s), in which case a specific solution can be found.

Example 2.1: Show that $y = x^2$ is a solution of $xy' = 2y$. Are there any other functions that solve this differential equation?

Solution: The derivative is $y' = 2x$ and by substitution, we have

$$x(2x) = 2(x^2)$$

$$2x^2 = 2x^2, \quad \text{true.}$$

Note that $y = 3x^2$ is also a solution. Its derivative is $y' = 6x$, and we have

$$x(6x) = 2(3x^2)$$
$$6x^2 = 6x^2, \quad \text{true.}$$

Generalizing, it appears any function of the form $y = Cx^2$ is a solution. The derivative would be $2Cx$, and we have

$$x(2Cx) = 2(Cx^2)$$
$$2Cx^2 = 2Cx^2, \quad \text{true.}$$

This illustrates that there are infinitely many solutions of this differential equations.

Example 2.2: Given that $y = Cx^2$ is a set of solutions of the differential equation $xy' = 2y$, find the specific solution satisfying the initial condition $y(2) = 3$.

Solution: From the previous example, we showed that $y = Cx^2$ is a set of functions that solve this differential equation. To find C, we evaluate $x = 2$ and $y = 3$ into the solution, and solve for C:

$$3 = C(2)^2$$
$$3 = 4C$$
$$C = \frac{3}{4}.$$

Thus, the specific solution that solves the differential equation and satisfies the initial condition is

$$y = \frac{3}{4}x^2.$$

Problems that include initial condition(s) are called **Initial Value Problems**, abbreviated **IVP**.

Example 2.3: Show that $y = e^{4x}$ is a solution of $y'' - y' - 12y = 0$. Are there other solutions?

Solution: The derivatives are $y' = 4e^{4x}$ and $y'' = 16e^{4x}$. Substituting, we have

$$\overbrace{(16e^{4x})}^{y''} - \overbrace{(4e^{4x})}^{y'} - 12\overbrace{(e^{4x})}^{y} = e^{4x}(16 - 4 - 12) = 0.$$

Another possible solution is $y = 10e^{4x}$:

$$(160e^{4x}) - (40e^{4x}) - 12(10e^{4x}) = e^{4x}(160 - 40 - 120) = 0.$$

In general, $y = Ce^{4x}$ forms a set of solutions of the differential equation. The set of functions $y = Ce^{-3x}$ also solve the differential equation (you verify). Their sums also are solutions. In this case, the general solution set is

$$y = C_1 e^{4x} + C_2 e^{-3x},$$

where C_1 and C_2 are two constants, not necessarily of the same value. Later, we will learn methods to finding these solutions and if given initial conditions, the values for the constants.

Example 2.4: What is a solution of $y' = 6x$? Are there other solutions? Generalize the solution set.

Solution: We rewrite the differential equation as $\frac{dy}{dx} = 6x$, so that $dy = 6x\,dx$. Then we integrate both sides:

$$\int dy = \int 6x\,dx$$

$$y = 3x^2 + C.$$

A possible solution is $y = 3x^2$. Another one is $y = 3x^2 - 4$, and so on. Note that the constant of integration C will disappear upon taking the derivative. This is an example of a simple separable differential equation. In fact, this is the type of problem commonly seen in an introductory calculus course when learning antidifferentiation and indefinite integrals. These are differential equations too!

Example 2.5: What is a solution of $y' = y$? Are there other solutions? Generalize the solution set.

Solution: The differential equation asks "what function y is equal to its derivative y'?". From calculus, we know that $y = e^x$ is a solution, which is easily verified. Are there other solutions?

Does $y = e^x + C$ form a set of solutions? Let's check. We have $y' = e^x$, so substituting, we have

$$\underbrace{e^x}_{y'} = \underbrace{e^x + C}_{y}.$$

This is not true if C is non-zero. Thus, $y = e^x + C$ does not form a set of solutions.

Does $y = Ce^x$ form a set of solutions? Let's check. We have $y' = Ce^x$, so substituting, we have

$$Ce^x = Ce^x.$$

This is true for all real numbers C. Thus, $y = Ce^x$ forms a set of solutions.

Example 2.6: What is a solution of $y'' = -y$? Are there other solutions? Generalize the solution set.

Solution: The differential equation asks "what function y, when negated, is equal to its second derivative y''?" From calculus, we know that the functions $y = \cos x$ and $y = \sin x$ fit this description:

$$y = \cos x, \; y' = -\sin x, \; y'' = -\cos x, \; \text{so} \; \underbrace{(-\cos x)}_{y''} = -\underbrace{(\cos x)}_{y}.$$

$$y = \sin x, \; y' = \cos x, \; y'' = -\sin x, \; \text{so} \; \underbrace{(-\sin x)}_{y''} = -\underbrace{(\sin x)}_{y}.$$

The general solution is

$$y = C_1 \cos x + C_2 \sin x.$$

> How do we know whether to add the constant C or attach it by multiplication to the function? As we discuss the various solution methods, this question will answer itself, in the sense that the steps, if done correctly, will naturally place C in its proper place.

Section 3
Classification of Differential Equations

Differential equations are classified according to their characteristics:

Ordinary: the terms of the differential equation are a function, $y = f(x)$, and/or at least one of its derivatives, $y' = f'(x)$, $y'' = f''(x)$ and so on, in the equation. There is just one independent variable. Such differential equations are called **Ordinary Differential Equations** and are abbreviated **ODE**. Differential equations involving partial derivatives are called Partial Differential Equations and are labeled PDE.

Every differential equation in this book is ordinary.

Order of a derivative: The first derivative of a function is considered a first-order derivative. The second derivative is called a second-order derivative, and so on. Higher-order derivatives are labeled $y^{(n)} = f^{(n)}(x)$ rather than writing out n prime symbols. The function $y = f(x)$ is considered zero-ordered. Differential equations are at minimum first-ordered.

Order of a differential equation: The highest order of the derivative present in the differential equation. Ordinary differential equations are usually written in terms of descending order.

Linearity: A differential equation is linear if it can be written in the form

$$a_n(x)y^{(n)} + a_{n-1}(x)y^{(n-1)} + \cdots + a_1(x)y' + a_0(x)y = g(x).$$

The orders of the derivatives of y must be non-negative integers (it does not make sense to discuss a half-ordered derivative, for example). The powers of y and its derivatives must be 1. For example, an expression such as $(y'')^4$ is a second-order derivative raised to the 4th power or degree. The coefficients $a_n(x)$ represent functions of x, possibly constants, that are attached to y and its derivatives by multiplication. The term $g(x)$, called a **forcing function**, is not attached to y or its derivatives by multiplication and may be a function of x, or just a constant, possibly 0. If the forcing function is 0, then the differential equation is **homogeneous**. Otherwise, it is **non-homogeneous**. The terms *forcing function*, *homogeneous* and *non-homogeneous* only apply to linear differential equations.

Autonomous: A differential equation is autonomous if it does not contain the independent variable anywhere in its simplified form. Autonomous differential equations may or may not be linear.

Example 3.1: Classify the differential equation: $y'' + 2xy = 3$.

Solution:

It is second order.
It is linear.
It is not homogeneous. The forcing function is $g(x) = 3$.
It is not autonomous. It contains the independent variable x in its simplified form.

Example 3.2: Classify the differential equation: $y^{(3)} + y' + 4y = 0$.

Solution:

It is third order.
It is linear.
It is homogeneous.
It is autonomous.

Example 3.3: Classify the differential equation: $y^{(4)} + x^2 y'' + \frac{1}{y} = 0$.

Solution:

It is fourth order.
It is not linear. The term $\frac{1}{y}$ has "order" -1, which, for linearity, is not allowed.
Since it is not linear, homogeneity is not relevant.
It is not autonomous.

Example 3.4: Classify the differential equation: $\frac{dy}{dx} = y^2$.

Solution:

It is first order.
It is not linear. The function y is raised to the second power (note: this is different from being of order 2)
Homogeneity is not relevant.
It is autonomous because it can be written as $y' = y^2$.

Example 3.5: Classify the differential equation: $(y'')^3 + y + x = 1$.

Solution:

It is second order.
It is not linear. The second derivative is raised to the power of 3.
Homogeneity is not relevant.
It is not autonomous.

Example 3.6: Classify the differential equation: $y^{(3)} + x^2 y'' + 2y - 3x = 0$.

Solution:

It is third order.
It is linear. It can be rewritten $y^{(3)} + x^2 y'' + 2y = 3x$
It is not homogeneous. The forcing function is $g(x) = 3x$.
It is not autonomous.

Example 3.7: Classify the differential equation: $y''y + x^2 = 1$.

Solution:

It is second order.
It is not linear since y and its second derivative are attached by multiplication.
Homogeneity is not relevant.
It is not autonomous.

Section 4
Separation of Variables & Direction Fields

A **separable** first-order differential equation is one that can be written so that the independent variable terms (along with its differential) are collected to one side of the equal sign, and the dependent variable terms (and its differential) to the other. In such cases, always write $\frac{dy}{dx}$ in place of y'. This is called **Separation of Variables**. Once the factors have been separated, then integrate both sides with respect to the appropriate variable, and simply.

Example 4.1: Find the general solution of $y' = xy^2$.

Solution: This differential equation is separable, written as

$$\frac{dy}{y^2} = x\, dx, \qquad y \neq 0.$$

To solve, antidifferentiate both sides:

$$\int \frac{dy}{y^2} = \int x\, dx.$$

Both sides should have a constant of integration added, but since they can be combined into one constant, it is sufficient to write one constant C, and it does not matter on which side it is written. Since we plan to solve for y, we write the C on the side of the x variable.

$$-\frac{1}{y} = \frac{1}{2}x^2 + C.$$

Now, solve for y. We multiply both sides by -1. Since C is a general constant for now, it "absorbs" the -1 and remains written as a general constant C.

$$\frac{1}{y} = -\frac{1}{2}x^2 + C.$$

Reciprocate both sides:

$$y = \frac{1}{-\frac{1}{2}x^2 + C}.$$

This is the solution set (general solution) of the differential equation $y' = xy^2$. It can be simplified. Multiply both numerator and denominator by 2 to clear the internal fraction:

$$y = \frac{2}{-x^2 + C}.$$

The C absorbs the 2 and remains written as C. A slightly cleaner form is

$$y = \frac{2}{C - x^2}, \qquad x \neq \pm\sqrt{C}, y \neq 0.$$

Always note any restrictions on the variable(s).

Let's discuss the intervals over which a specific solution of the differential equation may exist. Since $x \neq \pm\sqrt{C}$, this divides the Real line into three non-overlapping intervals: $x < -\sqrt{C}$, $-\sqrt{C} < x < \sqrt{C}$, and $x > \sqrt{C}$. To determine which interval "contains" (is the domain of) the specific solution, we need an initial condition, $y(x_0) = y_0$. We then determine C and then compare the location of the initial condition with the bound(s) of the three possible intervals and choose the correct one. Furthermore, in the case where there may be a restriction on the dependent variable, it will divide that variable's axis accordingly. Since $y \neq 0$, then the solutions will either have (at most) $y > 0$ or $y < 0$ as the range.

For example, if the initial condition is $y(1) = 4$, then we conclude that, at most, the range is $y > 0$. In fact, the range of y can be further refined once we know more about the specific solution.

Given the initial condition $y(1) = 4$ and evaluating into the general solution form, we find that $C = 1.5$, so the specific solution is

$$y = \frac{2}{1.5 - x^2}.$$

The three possible domains for x are $(-\infty, -\sqrt{1.5})$, $(-\sqrt{1.5}, \sqrt{1.5})$, or $(\sqrt{1.5}, \infty)$. Since $x_0 = 1$ is in the interval $(-\sqrt{1.5}, \sqrt{1.5})$, then this is the domain for the solution curve that passes through $(1, 4)$.

Regarding the range, using calculus, we can find a critical value for $y = \frac{2}{1.5 - x^2}$, which is $x = 0$. This lies in our solution's domain. This corresponds to the point $(0, 4/3)$, which is a minimum (you verify). The range of the solution is $[4/3, \infty)$.

Suppose the initial condition is $y(3) = -1$. Since $y_0 = -1$, then the potential range must be at most $y < 0$ (we will further refine this if necessary a few steps from now).

Evaluating this initial condition into the solution we find that $C = 7$ so that the specific solution is

$$y = \frac{2}{7-x^2}.$$

This means that the domain for the specific solution is within $(-\infty, -\sqrt{7})$, $(-\sqrt{7}, \sqrt{7})$, or $(\sqrt{7}, \infty)$. Since $x_0 = 3 > \sqrt{7}$, the specific solution has domain $(\sqrt{7}, \infty)$.

Differentiating y, the critical value occurs when $x = 0$, but this value lies outside the interval $(\sqrt{7}, \infty)$. Over the interval $(\sqrt{7}, \infty)$, the derivative $y' = \frac{4x}{(7-x^2)^2}$ is always positive, and already knowing that $y < 0$ and that $y = 0$ cannot be crossed, we can conclude that the specific solution increases but approaches the x-axis ($y = 0$) asymptotically. This can also be verified by graphing the solution.

Direction Fields

The set of solutions to a first-order differential equation can be visualized using a **direction field**. For example, consider

$$y' = xy^2. \quad \text{(Example 4.1)}$$

For each ordered pair in the plane, the value for y' is calculated and interpreted as a slope at that point. For example, at the ordered pair (2,3), we have

$$y' = (2)(3)^2 = 18.$$

Thus, at the point (2,3), a small arrow (vector) of slope 18 is sketched. The vector always points in the direction of increasing x. This is done for "all" points in the plane, and the result is the direction field for this differential equation. Direction fields are extremely time consuming to sketch by hand. Instead, graphing utilities or online applets are used.

Recall that the solution to this differential equation in Example 4.1 is

$$y = \frac{2}{C-x^2}, \quad x \neq \pm\sqrt{C}, y \neq 0.$$

Solution curves are inferred by starting at a given point and following the arrows. It is permissible to sketch the curve "in reverse" from the initial condition to get a more complete image of the curve. Note that none of the curves appear to cross or touch $y = 0$. This was a restriction that was noted in the example. The curves exist entire above the line $y = 0$, or entirely below it, depending on the initial condition. The direction field for $y' = xy^2$ (Example 4.1) is:

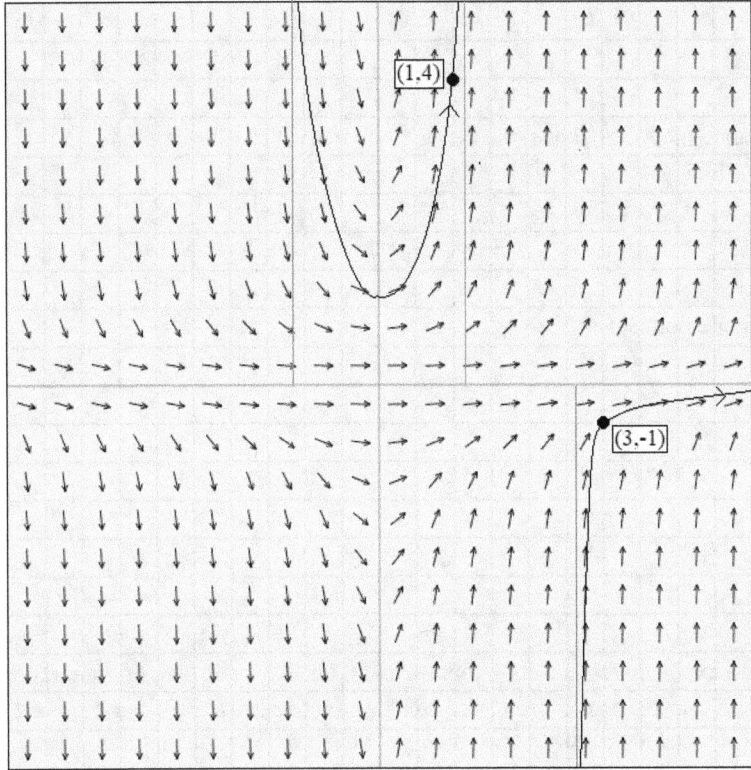

Scale: each gridline is 0.5 unit.　　Courtesy https://aeb019.hosted.uark.edu/dfield.html.

The two solution curves for $y = \frac{2}{1.5-x^2}$ corresponding to the initial condition (1,4) and for $y = \frac{2}{7-x^2}$ corresponding to (3, −1) are shown above. Each are shown with their respective bounds on x sketched in as vertical asymptotes.

Something interesting happens with the specific solutions of $y' = xy^2$: When the initial condition lies above the x-axis (that is, $y_0 > 0$), then the domain of the specific solution will always be within bounds $-\sqrt{C} < x < \sqrt{C}$, once C has been determined. And if the initial condition lies below the x-axis ($y_0 < 0$), then the domain of the specific solution will always be either $x < -\sqrt{C}$ or $x > \sqrt{C}$.

Example 4.2: Find the general solution of $y' = x + xy$.

Solution: Rewrite y' in differential form:

$$\frac{dy}{dx} = x + xy.$$

Factor:

$$\frac{dy}{dx} = x(1 + y).$$

Separate the variables including the individual differentials. Note any restrictions on any variables.

$$\frac{dy}{1+y} = x\,dx, \qquad y \neq -1.$$

Integrate both sides:

$$\int \frac{dy}{1+y} = \int x\,dx$$

$$\ln|1+y| = \frac{1}{2}x^2 + C.$$

Rewrite in base-e form:

$$|1+y| = e^{0.5x^2+C}.$$

This next step features an important detail. The constant of integration is currently in the exponent. Using rules of exponents, $e^{0.5x^2+C} = e^{0.5x^2}e^C$, but since C is a generic constant, so will e^C. Thus, we can write C again to represent this generic constant. The C now is connected to the other factor by multiplication:

$$|1+y| = Ce^{0.5x^2}$$

Removing the absolute value means to consider the $\pm$ cases opposite the equal sign. But as before, the C "absorbs" the signs for now. If there happens to be an initial condition, then C can be determined and it may be positive or negative.

$$1+y = \pm Ce^{0.5x^2}$$
$$1+y = Ce^{0.5x^2} \quad (\pm C = C)$$

Thus, $y = Ce^{0.5x^2} - 1$ is the general solution of $y' = x + xy$. In this case, there are no restrictions on x so that the solution domain will be the entire Real line. The range of a solution will be at most $y < -1$ or $y > -1$.

The direction field for $y' = x + xy$ is:

Scale: each gridline is 0.5 unit.   Courtesy https://aeb019.hosted.uark.edu/dfield.html.

There are no restrictions on x. The solution curves will be defined for all Real numbers.

Observe that $y = -1$ appears here as a bound between the set of solution curves above and below it. It is a barrier over which no solution curve can pass through. During the solution process, the restriction $y \neq -1$ appeared early (when the expression $y + 1$ appeared in a denominator). However, what happens when $y = -1$? Then $y' = 0$ and upon substitution into the original differential equation, we get $0 = x + x(-1)$ which is true for all x. The "solution" here is just the horizontal line $y = -1$. Such solution are called **trivial solutions** because they offer no insight to the general form of the set of solution curves. They are essentially ignored.

Example 4.3: Find the solution of $(1 + x^2)y' = \frac{1}{2y}$, where $y(1) = 3$.

Solution: This differential equation is separable, written as

$$2y\, dy = \frac{dx}{1 + x^2}.$$

To solve, antidifferentiate both sides:

$$\int 2y\, dy = \int \frac{dx}{1 + x^2}$$

$$y^2 = \tan^{-1} x + C$$

$$y = \pm\sqrt{\tan^{-1} x + C}.$$

To determine C, we use the initial condition, $y(1) = 3$. Since $y > 0$ in this case, only the positive square root need be considered:

$$3 = \sqrt{\tan^{-1} 1 + C}$$

$$9 = \frac{\pi}{4} + C \quad \left(\text{Since } \tan\frac{\pi}{4} = 1\right)$$

$$C = 9 - \frac{\pi}{4}.$$

The specific solution is

$$y = \sqrt{\tan^{-1} x + 9 - \frac{\pi}{4}}.$$

Let's now discuss this specific solution's domain and range. The domain for $\tan^{-1} x$ is all Real numbers. But since the expression is within a radical, we need to ensure that $\tan^{-1} x + 9 - \pi/4 \geq 0$. The range for $\tan^{-1} x$ is $(-\pi/2, \pi/2)$, so we study the two extremes. When $y \to -\pi/2$ (this happens when $x \to -\infty$, and furthermore, y approaches $-\pi/2$ from above, meaning $y > -\pi/2$ during this "approach"), we have within in the radical (as a limit) $-\pi/2 + 9 - \pi/4 \approx 6.6438\ldots$.

A similar process shows that as $x \to \infty$, then $y \to \pi/2$ from below. Within the radical, we have (as a limit) $\pi/2 + 9 - \pi/4 \approx 9.7854\ldots$. Taking square roots, we find that $\sqrt{6.6438} \approx 2.5775\ldots$ and that $\sqrt{9.7854} \approx 3.1281\ldots$. Note that $y_0 = 3$ lies within the interval $2.5775\ldots < y < 3.1281\ldots$. This is the range for the specific solution. The expression $\tan^{-1} x + 9 - \pi/4$ is never negative.

Its specific solution is graphed below, shown with the range for y:

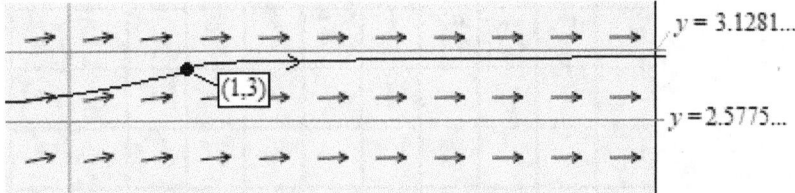

Scale: each gridline is 0.5 unit.   Courtesy https://aeb019.hosted.uark.edu/dfield.html.

Section 5
First-Order Autonomous Differential Equations

These differential equations are often solved using Separation of Variables. They are useful for modeling growth behavior, where the rate of growth is proportional in some manner to the quantity present, usually via a proportion.

Example 5.1: The rate of change of a population of a city is proportional to the population itself. If the population in 2010 was 25,000, and in 2018 was 32,000, forecast the population in 2025.

Solution: Let $P(t)$ be the population after t years, where $t = 0$ represents the year 2010. Note that $P > 0$. Time t could be negative, for example, if we wanted to estimate the city's population in 2005. Translated, we obtain

"The rate of change of a population": $\quad\dfrac{dP}{dt}$

"is": $\quad =$

"proportional to the population itself": $\quad kP.$

The number k is the proportionality constant. It is not a constant of integration. Also, when $t = 0$, the population is 25,000, and when $t = 8$, the population is 32,000. Assembling this into an equation, we have

$$\frac{dP}{dt} = kP, \quad \text{where} \quad P(0) = 25{,}000 \quad \text{and} \quad P(8) = 32{,}000.$$

Separate the variables. Note that it makes sense for $P > 0$ in this example:

$$\frac{dP}{P} = k\, dt.$$

Integrate and solve for P:

$$\int \frac{dP}{P} = \int k\, dt$$

$$\ln P = kt + C$$

$$P = e^{kt+C}$$

$$P = e^{kt} e^{C}$$

$$P(t) = Ce^{kt}.$$

Use the ordered pair (0, 25000) to determine C:

$$25{,}000 = Ce^{0t}$$

$$C = 25{,}000.$$

Thus, we now have

$$P(t) = 25{,}000 e^{kt}.$$

Use the ordered pair (8, 32000) to determine k:

$$32{,}000 = 25{,}000 e^{k(8)}$$

$$\frac{32}{25} = e^{8k}$$

$$\ln\left(\frac{32}{25}\right) = 8k$$

$$k = \frac{1}{8}\ln\left(\frac{32}{25}\right) \approx 0.031.$$

The specific solution that meets all stated conditions is

$$P(t) = 25{,}000 e^{0.031 t}.$$

In 2025, $t = 15$, so we have

$$P(15) = 25{,}000 e^{0.031(15)} \approx 39{,}800.$$

In 2025, there will be about 39,800 people in the city.

Example 5.2: The rate of change in the value of a stock is inversely proportional to the square of the value of that stock. If the stock's value was $20 at noon, and was $23 at 3 p.m., what is the stock's value at 5 p.m.?

Solution: Let $V(t)$ be the value, in dollars, of the stock t hours after noon (when $t = 0$). Translating into mathematics, we have

"The rate of change in the value..."　　　$\dfrac{dV}{dt}$

"is"　　　$=$

"inversely proportional to the square of the value..."　　　$\dfrac{k}{V^2}$

The known conditions are $V(0) = \$20$ and $V(3) = \$23$. Thus, the differential equation that models this growth is

$$\frac{dV}{dt} = \frac{k}{V^2} \quad \text{where} \quad V(0) = 20 \quad \text{and} \quad V(3) = 23.$$

Separate the variables:

$$V^2 \, dV = k \, dt$$

Integrate:

$$\int V^2 \, dV = \int k \, dt, \quad \text{which gives} \quad \frac{1}{3}V^3 = kt + C.$$

Multiply both sides by 3 to clear fractions:

$$V^3 = 3kt + C.$$

The constant of integration C is a generic, so $3C$ is the same as writing C. Take the cube root:

$$V(t) = \sqrt[3]{3kt + C}.$$

This is the general model that governs the stock's value.

To find C, use the condition, $V(0) = 20$:

$$20 = \sqrt[3]{3k(0) + C}$$
$$20 = \sqrt[3]{C}$$
$$C = 20^3 = 8,000.$$

We now have
$$V(t) = \sqrt[3]{3kt + 8{,}000}.$$

To find k, use the other condition, $V(3) = 23$:
$$23 = \sqrt[3]{3k(3) + 8{,}000}.$$
$$23^3 = 9k + 8{,}000$$
$$12{,}167 = 9k + 8{,}000$$
$$4{,}167 = 9k$$
$$k = \frac{4{,}167}{9} = 463.$$

The specific solution is now
$$V(t) = \sqrt[3]{3(463)t + 8{,}000} = \sqrt[3]{1{,}389t + 8{,}000}.$$

In this case, it's permissible to combine the factors 3 and k in front of the t. It won't affect the later calculation.

The stock's value at 5 p.m. means $t = 5$:
$$V(5) = \sqrt[3]{1{,}389(5) + 8{,}000} \approx \$24.63.$$

Example 5.3: Find the general solution of $y' = y^2 + y$.

Solution: Separating variables gives
$$\frac{dy}{y^2 + y} = dx, \quad y \neq 0, -1.$$

Before antidifferentiating the left side, the denominator needs to be written as the sum of smaller fractions, using a process called partial fraction decomposition:
$$\frac{1}{y^2 + y} = \frac{1}{y(y+1)} = \frac{A}{y} + \frac{B}{y+1},$$

where A and B are the unknown numerators of the smaller fractions. The two fractions are then recomposed by finding the common denominator:

$$\frac{1}{y(y+1)} = \frac{A(y+1) + By}{y(y+1)}.$$

The numerators are now compared. On the right side, clear parentheses and reorder the terms according to powers of y:

$$1 = (A + B)y + A.$$

By viewing the left side as $0y + 1$, relate the expressions on the right to those on the left. Thus, $A + B = 0$ and $A = 1$. This forces $B = -1$. The partial fraction decomposition is complete, and we have

$$\frac{1}{y^2 + y} = \frac{1}{y} - \frac{1}{y+1}.$$

This is now in a form to be antidifferentiated:

$$\int \frac{dy}{y^2 + y} = \int dx$$

$$\int \left(\frac{1}{y} - \frac{1}{y+1}\right) dy = \int dx$$

$$\ln y - \ln(y+1) = x + C.$$

To solve for y, use the logarithm property $\ln a - \ln b = \ln \frac{a}{b}$:

$$\ln \left(\frac{y}{y+1}\right) = x + C.$$

This is rewritten using base-e notation. Note that $e^{x+C} = e^x e^C = Ce^x$. Then y is isolated through algebra:

$$\frac{y}{y+1} = Ce^x$$

$$y = Ce^x(y+1)$$

$$y = Cye^x + Ce^x$$

$$y - Cye^x = Ce^x$$

$$y(1 - Ce^x) = Ce^x$$

$$y = \frac{Ce^x}{1 - Ce^x}.$$

While this is correct, a simpler form can be found by dividing the numerator and denominator by Ce^x:

$$\frac{Ce^x/Ce^x}{(1-Ce^x)/Ce^x} = \frac{1}{\left(\frac{1}{Ce^x}\right)-1}.$$

The expression $\frac{1}{Ce^x} = Ce^{-x}$ by treating $\frac{1}{C}$ as "new" C. The simplified form of the general solution is

$$y = \frac{1}{Ce^{-x}-1}, \quad x \neq \ln C.$$

Let's explore the behavior of the solution curves, each dependent on an initial condition (a, b). Direct evaluation of the initial condition in the general solution shows that $C = \frac{1+b}{b}e^a$. When $b < -1$ or $b > 0$, then C is positive, and when $-1 < b < 0$, then C is negative.

The range of a solution curve will be in the interval $y < -1$, $-1 < y < 0$ or $y > 0$, where C is positive when $y < -1$ or $y > 0$, and C is negative when $-1 < y < 0$. Three cases emerge:

If $y < -1$, then C is positive, but the denominator $Ce^{-x} - 1$ must be negative for the whole expression $1/(Ce^{-x} - 1)$ to be negative, thus $Ce^{-x} - 1 < 0$. Isolating x, we find that the domain will be in the interval $x > \ln C$.

If $-1 < y < 0$, then the denominator must also be negative, but that since C is negative also, the expression $\ln C$ is not defined. In other words, the domain of x is the whole Real line.

If $y > 0$, then C is positive, and the denominator must be positive for the whole expression $1/(Ce^{-x} - 1)$ to be positive, thus $Ce^{-x} - 1 > 0$. Isolating x, we find that the domain will be in the interval $x < \ln C$.

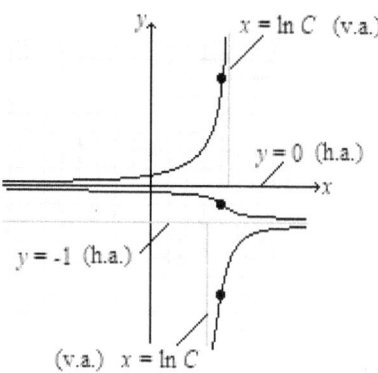

In the image, the asymptotes show how the xy-plane is divided into potential solution regions. The initial conditions for this illustration are $(2,3)$, $(2, -1/2)$ and $(2, -3)$. For the initial condition $(2,3)$, $C = \frac{4}{3}e^2$ so that $\ln\left(\frac{4}{3}e^2\right) \approx 2.288$ is the upper bound of the domain, and for $(2, -3)$, $C = \frac{2}{3}e^2$ so that $\ln\left(\frac{2}{3}e^2\right) \approx 1.595$ is the lower bound of the domain. Each is a vertical asymptote for its respective solution curve.

Direction Fields for First-Order Autonomous Equations

For an autonomous differential equation, the x is absent, so the slopes only depend on the y value at that point. For each y-value, the slope at each point is the same, going across horizontally. For example, the direction field for $y' = y$ is

Scale: each gridline is 0.5 unit. Courtesy https://aeb019.hosted.uark.edu/dfield.html.

The solution curves will trend away from the x-axis ($y = 0$), either entirely above the x-axis or entirely below it, depending on the initial condition. No solution curve for this differential equation will cross (or touch) the x-axis. Recall that the general solution for $y' = y$ is $y = Ce^x$. Picking any point in the plane (except along the x-axis), and following the arrows, one sees the familiar shape of the exponential function.

In an autonomous differential equation, the y values for which $y' = 0$ are called **equilibrium** solutions. On a direction field, equilibrium solutions will be where the slope lines are horizontal, reading across left to right. There are three types of equilibrium solutions:

If the solution curves trend away from the equilibrium both above and below as x increases, it is an **unstable** equilibrium.

If the solution curves trend toward the equilibrium asymptotically both above and below as x increases, it is a **stable** equilibrium.

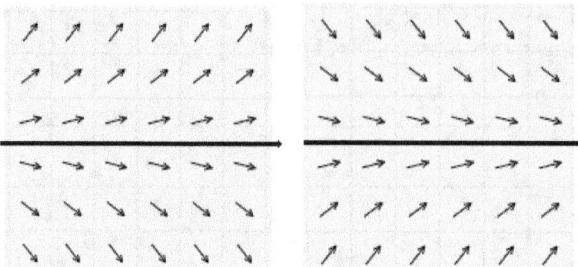

Left: example of an unstable equilibrium
Right: Example of a stable equilibrium

If the solution curves trend away from the equilibrium on one side of the equilibrium, and trend toward the equilibrium on the other side as x increases, it is a **semistable** equilibrium.

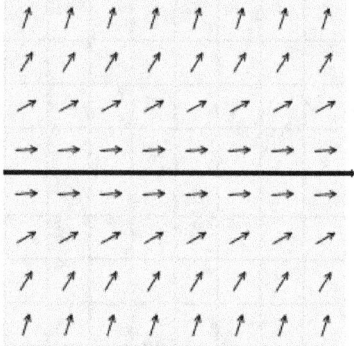

Example of a semistable equilibrium.
The curves trend away from equilibrium when $y > 0$, and trend toward equilibrium when $y < 0$.

Example 5.4: Given $y' = y^3 - 9y$. Find all equilibrium solutions, and determine if they are stable, unstable or semistable.

Solution: The equilibrium points are where $y' = y^3 - 9y = 0$. Factoring, we have $y(y+3)(y-3) = 0$, so the equilibrium solutions are where $y = 0, y = 3$, and $y = -3$. This divides the y-axis into four intervals. A value is chosen within each interval and evaluated to determine the sign of the slope:

$y > 3$: Choose $y = 4$. Thus, $y' = 4^3 - 9(4) > 0$.
$0 < y < 3$: Choose $y = 1$. Thus, $y' = 1^3 - 9(1) < 0$.
$-3 < y < 0$: Choose $y = -1$. Thus, $y' = (-1)^3 - 9(-1) > 0$.
$y < -3$: Choose $y = -4$. Thus, $y' = (-4)^3 - 9(-4) < 0$.

Curves above $y = 3$ slope upward, away from $y = 3$, and curves between $y = 0$ and $y = 3$ will curve down, also away from $y = 3$. So $y = 3$ is unstable. By similar reasoning, $y = 0$ is stable and $y = -3$ is unstable.

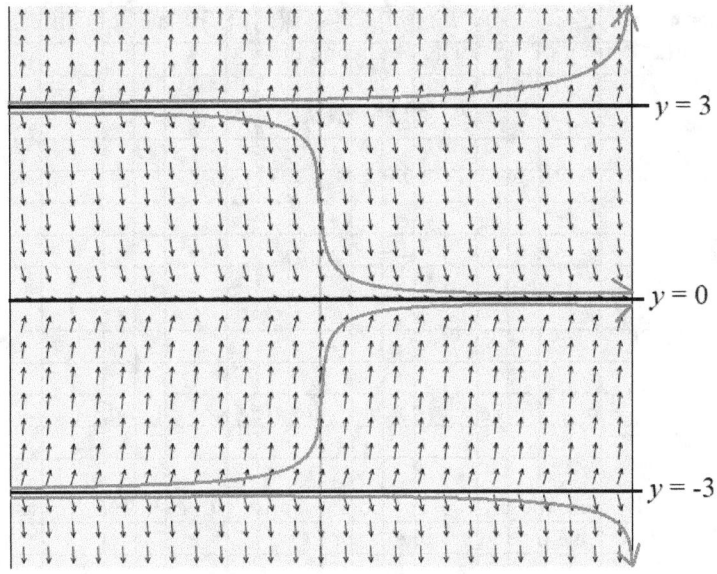

Direction field showing stable equilibrium at $y = 0$, and unstable equilibrium at $y = \pm 3$.
Courtesy https://aeb019.hosted.uark.edu/dfield.html.

Section 6
Integration Factors

There is a process by which most first-order linear differential equations can be solved. This uses an **integration factor**, denoted $\mu(x)$ (Greek letter "mu").

The differential equation must be in the form

$$y' + f(x)y = g(x).$$

This type of differential equation is sometimes solvable by separation of variables. However, if it is not possible to separate the variables, the integration factor method should be used.

Starting with $y' + f(x)y = g(x)$, multiply both sides by $\mu(x)$, which is currently undetermined:

$$\mu(x)y' + \mu(x)f(x)y = \mu(x)g(x).$$

The left side "looks like" a product-rule derivative of $(\mu(x)y)$:

$$(\mu(x)y)' = \mu(x)y' + \mu'(x)y.$$

Comparing $\mu(x)y' + \mu'(x)y$ with $\mu(x)y' + \mu(x)f(x)y$, the only difference is the factor(s) in front of the y. Thus, equating them gives

$$\mu'(x) = \mu(x)f(x).$$

This is a separable differential equation in terms of μ:

$$\frac{d\mu}{\mu(x)} = f(x)\,dx.$$

Integrating both sides, we have

$$\int \frac{d\mu}{\mu(x)} = \int f(x)\,dx, \quad \text{so that} \quad \ln\mu(x) = \int f(x)\,dx + C.$$

Here, we only need one form of the antiderivative, so let $C = 0$. Rewriting as base-e, we now have the integrating factor $\mu(x)$:

$$\mu(x) = e^{\int f(x)\,dx}.$$

Now that we know $\mu(x)$, return to the statement from above,

$$\mu(x)y' + \mu(x)f(x)y = \mu(x)g(x).$$

The left side is a product rule derivative, $(\mu(x)y)'$. So, we have

$$(\mu(x)y)' = \mu(x)g(x).$$

Integrating, we have

$$\int (\mu(x)y)' = \int \mu(x)g(x)\,dx + C.$$

Since $\int (\mu(x)y)' = \mu(x)y$, we have:

$$\mu(x)y = \int \mu(x)g(x)\,dx + C.$$

Solving for y gives us the formula for the solution:

$$y = \frac{\int \mu(x) g(x) \, dx + C}{\mu(x)}.$$

You may memorize this as a formula. Don't forget the "+ C" in the numerator.

Example 6.1: Find the general solution of $y' + \frac{2}{x} y = x$. Note that $f(x) = \frac{2}{x}$ and $g(x) = x$ and that $x \neq 0$.

Solution: First, we find $\mu(x)$:

$$\mu(x) = e^{\int \frac{2}{x} dx} = e^{2 \ln x} = e^{\ln x^2} = x^2.$$

Now, use the formula $y = \frac{\int \mu(x) g(x) \, dx + C}{\mu(x)}$:

$$y = \frac{\int x^2 x \, dx + C}{x^2} = \frac{\int x^3 \, dx + C}{x^2} = \frac{\frac{1}{4} x^4 + C}{x^2} = \frac{1}{4} x^2 + C x^{-2}.$$

Thus, $y = \frac{1}{4} x^2 + C x^{-2}$ is the general solution of $y' + \frac{2}{x} y = x$. The restriction $x \neq 0$ still remains in effect.

Check: First, differentiate y:

$$y' = \frac{1}{2} x - 2 C x^{-3}$$

Note: The C does not absorb the 2 in this step. It must remain written as $-2C$. This is because whatever C is in the solution, $-2C$ will appear in its derivative.

Now insert y' and y into the differential equation and simplify:

$$\left(\frac{1}{2} x - 2 C x^{-3} \right) + \frac{2}{x} \left(\frac{1}{4} x^2 + C x^{-2} \right) = x$$

$$\left(\frac{1}{2} x - 2 C x^{-3} \right) + \left(\frac{1}{2} x + 2 C x^{-3} \right) = x$$

$$\left(\frac{1}{2} x + \frac{1}{2} x \right) + (-2 C x^{-3} + 2 C x^{-3}) = x$$

$$x + 0 = x.$$

Example 6.2: Solve the IVP: $y' + 2xy = 3x$, $y(0) = 5$.

Solution: Find $\mu(x)$:

$$\mu(x) = e^{\int 2x \, dx} = e^{x^2}.$$

Now find y using the formula $y = \frac{\int \mu(x) g(x) \, dx + C}{\mu(x)}$:

$$y = \frac{\int (e^{x^2})(3x) \, dx + C}{e^{x^2}} = \frac{\int 3x e^{x^2} \, dx + C}{e^{x^2}} = \frac{\frac{3}{2} e^{x^2} + C}{e^{x^2}} = \frac{3}{2} + C e^{-x^2}.$$

To find the specific solution, use the initial condition $y(0) = 5$:

$$5 = \frac{3}{2} + C e^{-(0)^2}$$

$$5 = \frac{3}{2} + C$$

$$C = \frac{7}{2}.$$

The specific solution is

$$y = \frac{3}{2} + \frac{7}{2} e^{-x^2}.$$

There are no restrictions on x. Furthermore, since $0 < e^{-x^2} \leq 1$, the range is $\frac{3}{2} < y \leq 5$.

Example 6.3: Use an integration factor to solve $\frac{dy}{dx} = x + xy$. (This was solved previously using separation of variables; See Example 4.2)

Solution: Rewrite in the form $y' + f(x) y = g(x)$:

$$y' - xy = x.$$

The integration factor is

$$\mu(x) = e^{\int -x \, dx} = e^{-0.5 x^2}.$$

34

The solution is

$$y = \frac{\int xe^{-0.5x^2} dx + C}{e^{-0.5x^2}} = \frac{-e^{-0.5x^2} + C}{e^{-0.5x^2}} = -1 + Ce^{0.5x^2}.$$

There are no restrictions on x.

This is the same solution as found in Example 4.2. Both methods worked here, so it is your choice whether to try one over another first. Generally, separation of variables is faster, but the method of integration factors solves more differential equations, especially those that are not separable.

Example 6.4: Use an integration factor to solve $y' + 2xy = 1$.

Solution: First, note that this differential equation is not separable.

The integration factor is

$$\mu(x) = e^{\int 2x\, dx} = e^{x^2}.$$

Using the formula for the solution, we have

$$y = \frac{\int e^{x^2} dx + C}{e^{x^2}}.$$

This is as far as we can go, the integral $\int e^{x^2} dx$ is not expressible in common functions. This does not mean the differential equation does not have a solution. It does. It's the equation above. The best we can do is leave it in the integral form.

In cases like this, a numerical method may be used to find solution curves, if not an actual formula using common functions. This example is revisited in Example 9.2. A series can also be used to express terms that approximate a solution to ever-increasing precision. This is revisited in the appendix under the Series Solutions section.

Section 7
Bernoulli Equations

A **Bernoulli Equation** has the form

$$y' + P(x)y = Q(x)y^n.$$

Non-linear differential equations are often difficult to solve, but in this case, there does exist a general solution algorithm. Note that when $n = 0$ or $n = 1$, the above equation is linear and can be solved by separation of variables or with an integration factor. Also note that if $P(x)$ and $Q(x)$ are both constants, then the equation is autonomous and can be solved by separation of variables, although the algebra can be challenging.

Divide through by y^n:

$$y^{-n}y' + P(x)y^{1-n} = Q(x).$$

Perform a variable change. Let $u = y^{1-n}$. Remember, both functions are with respect to x. Using the Chain Rule, the derivative of u with respect to x is:

$$\frac{du}{dx} = (1-n)y^{-n}\frac{dy}{dx}, \quad \text{or} \quad u' = (1-n)y^{-n}y'.$$

Note that $y^{-n}y' = \frac{1}{1-n}u'$. Thus, we now make the substitutions:

$$\underbrace{y^{-n}y'}_{\frac{1}{1-n}u'} + P(x)\underbrace{y^{1-n}}_{u} = Q(x).$$

This is a linear first-order differential equation in terms of u. Multiplying through by $1 - n$ clears fractions, and the general formula for determining u is

$$u' + (1-n)P(x)u = (1-n)Q(x).$$

Example 7.1: Find the general solution of $y' + xy = xy^2$.

Solution: Here, $n = 2$ and the conversion is $u = y^{-1}$. We don't need to perform all the steps above, as we can make substitutions in the formula to find u. However, once we find u, then we will need to rewrite it in terms of y, and this is where the conversion will be necessary. After substitutions are made, the differential equation in terms of u is

$$u' - xu = -x.$$

This is solved using an integration factor:

$$\mu(x) = e^{\int -x\, dx} = e^{-0.5x^2}.$$

The solution in terms of u is

$$u = \frac{\int (e^{-0.5x^2})(-x)\, dx + C}{e^{-0.5x^2}} = \frac{e^{-0.5x^2} + C}{e^{-0.5x^2}} = 1 + Ce^{0.5x^2}.$$

Since $u = y^{-1}$, then $y = u^{-1}$. Thus, the general solution is

$$y = \frac{1}{1 + Ce^{0.5x^2}}.$$

Check: The derivative of y with respect to x is

$$y' = \frac{-Cxe^{0.5x^2}}{(1 + Ce^{0.5x^2})^2}.$$

Substitute into the original differential equation:

$$\underbrace{\frac{-Cxe^{0.5x^2}}{(1 + Ce^{0.5x^2})^2}}_{y'} + x\underbrace{\left(\frac{1}{1 + Ce^{0.5x^2}}\right)}_{y} = x\underbrace{\left(\frac{1}{1 + Ce^{0.5x^2}}\right)^2}_{y^2}.$$

Multiply through by $(1 + Ce^{0.5x^2})^2$ to clear fractions:

$$-Cxe^{0.5x^2} + x(1 + Ce^{0.5x^2}) = x$$
$$-Cxe^{0.5x^2} + x + Cxe^{0.5x^2} = x, \quad \text{true.}$$

Example 7.2: Find the solution of $y' + \frac{2}{x}y = y^3$ where $y(1) = 3$. $(x \neq 0)$

Solution: We have $n = 3$ so that $u = y^{-2}$. After substitution, the differential equation in terms of u is

$$u' - 2\left(\frac{2}{x}\right)u = -2, \quad \text{or} \quad u' - \frac{4}{x}u = -2.$$

Solve this equation using an integration factor:

$$\mu(x) = e^{\int -\frac{4}{x}\, dx} = e^{-4\ln x} = x^{-4}.$$

The solution in terms of u is

$$u = \frac{\int (x^{-4})(-2)\,dx + C}{x^{-4}} = \frac{\frac{2}{3}x^{-3} + C}{x^{-4}} = \frac{2}{3}x + Cx^4.$$

Since $u = y^{-2}$, then $y = \frac{1}{\sqrt{u}} = u^{-1/2}$. The general solution in terms of y is

$$y = \pm\left(\frac{2}{3}x + Cx^4\right)^{-1/2}.$$

Since $y > 0$ in the initial condition, consider only the positive root. To find C, use the initial condition $y(1) = 3$:

$$3 = \left(\frac{2}{3}(1) + C(1)^4\right)^{-1/2}$$

$$3 = \left(\frac{2}{3} + C\right)^{-1/2}$$

$$\frac{1}{9} = \frac{2}{3} + C$$

$$C = -\frac{5}{9}.$$

The specific solution is

$$y = \left(\frac{2}{3}x - \frac{5}{9}x^4\right)^{-1/2} = \frac{3}{\sqrt{6x - 5x^4}}.$$

Note that the expression $6x - 5x^4$ cannot be negative (as it is within an even root) nor 0 (as it is in a denominator). Thus,

$$6x - 5x^4 > 0.$$

Factor:

$$x(6 - 5x^3) > 0.$$

Sign changes occur when $x = 0$ (which is outside of the domain) or $x = \sqrt[3]{6/5} \approx 1.0627$. The above expression is negative when $x < 0$ or $x > 1.0627$. The initial condition, (1,3), suggests that $y > 0$ when $x = 1$, so the interval of validity (the domain) for this solution is $0 < x < 1.0627$.

When $P(x)$ and $Q(x)$ are both constant, the differential equation is separable. The following differential equation was solved in Example 5.3 using separation of variables. Let's solve it again using the Bernoulli method and compare the processes.

Example 7.3: Find the general solution of $y' = y^2 + y$.

Solution: Rewrite as $y' - y = y^2$ and note that $P(x) = -1$ and $Q(x) = 1$. Also, note that $n = 2$ so that $u = y^{-1}$. After substitutions, the new differential equation in terms of u is:

$$u' + u = -1.$$

The integration factor is

$$\mu(x) = e^{\int dx} = e^x.$$

The solution in terms of u is

$$u = \frac{\int (e^x)(-1)\, dx + C}{e^x} = \frac{-e^x + C}{e^x} = -1 + Ce^{-x}.$$

Since $u = y^{-1}$, then $y = u^{-1}$, so the solution in terms of y is

$$y = (-1 + Ce^{-x})^{-1} = \frac{1}{Ce^{-x} - 1}.$$

This is the same result as found in Example 5.3. The method of separation of variables used in Example 5.3 was cumbersome due to the need to use partial fraction decomposition. This method is an alternative in such cases.

The value of n may not be an integer.

Example 7.4: Find the solution of $y' - 3y = \sqrt{y}$, $y \geq 0$.

Solution: We have $n = 1/2$ so that $u = y^{1/2}$. After substitution, the differential equation in terms of u is

$$u' - \frac{3}{2}u = \frac{1}{2}.$$

Solve this equation using an integration factor:

$$\mu(x) = e^{\int -(3/2)\, dx} = e^{-(3/2)x}.$$

The solution in terms of u is:

$$u = \frac{\int (e^{-(3/2)x})(0.5) + C}{e^{-(3/2)x}} = \frac{-\frac{1}{3} e^{-(3/2)x} + C}{e^{-(3/2)x}} = -\frac{1}{3} + C e^{(3/2)x}.$$

Since $u = y^{1/2}$, we have $y = u^2$. Therefore, the general solution is

$$y = \left(-\frac{1}{3} + C e^{(3/2)x}\right)^2.$$

Note that the requirement that $y \geq 0$ is preserved.

Section 8
Mixture Problems

A common application of first-ordered linear differential equations is mixture problems. These problems are helpful in understanding the solution processes, in a tangible sense. In the first example, the amount of mixture entering and leaving the tank is the same amount per unit of time.

Example 8.1: A tank contains 1,000 kg of salt suspended in 10,000 liters of water. A mixture of 2 kg of salt per 10 liters of water enters the tank at one end at a rate of 4 liters per minute. It is mixed with what is already in the tank, and at the other end, the mixture leaves the tank at 4 liters per minute. Find $Q(t)$, the amount of salt in the tank after t minutes and find the limiting amount of salt in the tank.

Solution: Since the amount of salt changes continuously over time, we need to look at the rate, $\frac{dQ}{dt}$. In general,

$$\frac{dQ}{dt} = (\text{rate in}) - (\text{rate out}).$$

Fleshed out more, the formula's structure becomes

$$\frac{dQ}{dt} = \left(\frac{kg}{liters}\right)\left(\frac{liters}{minute}\right) - \left(\frac{kg}{liters}\right)\left(\frac{liters}{minute}\right).$$

We fill in the empty slots:

$$\frac{dQ}{dt} = \left(\frac{2 \text{ kg}}{10 \text{ liters}}\right)\left(\frac{4 \text{ liters}}{1 \text{ minute}}\right) - \left(\frac{Q(t) \text{ kg}}{10{,}000 \text{ liters}}\right)\left(\frac{4 \text{ liters}}{1 \text{ minute}}\right)$$

Note that both rates simplify as kg/min. Also note that in the "rate out", the amount of salt varies, so we use $Q(t)$ in this position, as it is still yet unknown.

Simplified, this is a differential equation:

$$Q' = \frac{4}{5} - \frac{1}{2{,}500}Q, \quad Q(0) = 1{,}000, \quad t \geq 0.$$

In problems like these, decimal notation may be cleaner:

$$Q' = 0.8 - 0.0004Q, \quad Q(0) = 1{,}000.$$

Rearranged slightly, we have

$$Q' + 0.0004Q = 0.8, \quad Q(0) = 1{,}000.$$

This is solved using an integration factor:

$$\mu(t) = e^{\int 0.0004 \, dt} = e^{0.0004t}.$$

Using the solution formula, we have

$$Q(t) = \frac{\int (e^{0.0004t})(0.8) \, dt + C}{e^{0.0004t}}.$$

Now we solve:

$$Q(t) = \frac{2{,}000 e^{0.0004t} + C}{e^{0.0004t}} = 2{,}000 + Ce^{-0.0004t}.$$

Use the initial condition to find C:

$$1{,}000 = 2{,}000 + Ce^{0.0004(0)}$$
$$1{,}000 = 2{,}000 + C$$
$$C = -1{,}000.$$

The specific solution is

$$Q(t) = 2{,}000 - 1{,}000 e^{-0.0004t}.$$

As $t \to \infty$ (in other words, this process runs a very long time), the limit

$$\lim_{t \to \infty} e^{-0.0004t} = 0.$$

Thus, the limiting amount of salt in the tank is 2,000 kg. Is this plausible? The tank originally contained 1,000 kg of salt in 10,000 liters, a concentration of 10%, the same as 1 kg per 10 liters (remember, 1 liter of pure water at sea level is equivalent to 1 kg). The solution coming in is 2 kg per 10 liters, or a 20% concentration. Over time, this will slowly raise the overall concentration from 10% to 20%, so 20% of 10,000 is 2,000. This answer seems plausible.

In this example, the mixture entering the tank differs in rate to what's leaving the tank.

Example 8.2: A vat has a capacity of 15,000 liters. It initially contains 2,000 liters of water in which 50 g of sugar has been dissolved. A mixture of 1.5 g of sugar per liter of water comes into the vat at 5 liters per minute. It mixes with what's in the vat, and at the other end, the mixture exits at the rate of 3 liters per minute. Find $Q(t)$, the quantity (in g) of sugar at time t, and find the amount of sugar in the tank when the tank fills to capacity (at which the process stops).

Solution: We start with

$$\frac{dQ}{dt} = (\text{rate in}) - (\text{rate out}).$$

$$\frac{dQ}{dt} = \left(\frac{\text{grams}}{\text{liters}}\right)\left(\frac{\text{liters}}{\text{minute}}\right) - \left(\frac{\text{grams}}{\text{liters}}\right)\left(\frac{\text{liters}}{\text{minute}}\right).$$

When the values are entered, we have

$$\frac{dQ}{dt} = \left(\frac{1.5 \text{ grams}}{1 \text{ liters}}\right)\left(\frac{5 \text{ liters}}{1 \text{ minute}}\right) - \left(\frac{Q(t) \text{ grams}}{(2{,}000 + 2t) \text{ liters}}\right)\left(\frac{3 \text{ liters}}{1 \text{ minute}}\right)$$

Note that since the net change of mixture coming in is 2 liters more than the amount leaving, the number of liters in the tank is not constant. It increases by 2 liters every minute. Thus, the expression $2000 + 2t$ accounts for the increasing amount of overall mixture in the tank. Also, since the tank has a capacity of 15,000 liters, the tank will fill in

$$2000 + 2t = 15{,}000$$
$$2t = 13{,}000$$
$$t = 6{,}500 \text{ minutes.}$$

Thus, the domain for this problem is $0 \le t \le 6{,}500$.

Simplified and re-arranged slightly, the differential equation is

$$Q' + \frac{3}{2000 + 2t} Q = 7.5, \qquad Q(0) = 50, \qquad 0 \le t \le 6{,}500.$$

This is solved using an integration factor:

$$\mu(t) = e^{\int (3/(2000+2t))\, dt} = e^{1.5 \ln(2000 + 2t)} = (2000 + 2t)^{1.5}.$$

The solution is

$$Q(t) = \frac{\int (2000 + 2t)^{1.5} (7.5)\, dt + C}{(2000 + 2t)^{1.5}}$$

$$= \frac{1.5(2000 + 2t)^{2.5} + C}{(2000 + 2t)^{1.5}}$$

$$= 1.5(2000 + 2t) + C(2000 + 2t)^{-1.5}$$

$$= 3000 + 3t + C(2000 + 2t)^{-1.5}.$$

We use the initial condition to determine C:

$$50 = 3000 + 3(0) + C\bigl(2000 + 2(0)\bigr)^{-1.5}$$
$$50 = 3000 + C(2000)^{-1.5}$$
$$C = -2950(2000)^{1.5}.$$

There is no need to work out the value of C. This is sufficient. The solution is

$$Q(t) = 3000 + 3t \overbrace{-2950(2000)^{1.5}}^{C} (2000 + 2t)^{-1.5}.$$

The flow runs for $t = 6500$ minutes until the vat is filled. At that time, there is

$$Q(6500) = 3000 + 3(6500) - 2950(2000)^{1.5}\bigl(2000 + 2(6500)\bigr)^{-1.5}$$
$$Q(6500) = 3000 + 19500 - 2950(2000)^{1.5}(15000)^{-1.5}$$
$$Q(6500) \approx 22{,}356.4 \text{ g of salt}.$$

Is this plausible? Suppose the process ran forever (equal amounts coming in and leaving). A 15,000-liter vat of solution would have 1.5 g of salt per liter, or a maximum of $15{,}000(1.5) = 22{,}500$ g of salt. But because this process stops, we expect our figure to be a little below that maximum, so an answer of 22,356.4 g of salt is very plausible.

Example 8.3: A smoky room that measures 6 m wide by 7 m long by 3 m high contains 2 grams of smoke particles per cubic meter. The windows are opened so that fresh air comes in at 0.5 cubic meters per minute, mixes with the air in the room, and exits the other end at the same rate. How much time will pass until the room contains just 0.15 grams of smoke per cubic meter?

Solution: We start with the $\frac{dQ}{dt}$ = rate in − rate out form:

$$\frac{dQ}{dt} = \left(\frac{0 \text{ g}}{1 \text{ m}^3}\right)\left(\frac{0.5 \text{ m}^3}{1 \text{ min.}}\right) - \left(\frac{Q(t) \text{ g}}{126 \text{ m}^3}\right)\left(\frac{0.5 \text{ m}^3}{1 \text{ min.}}\right), \quad Q(0) = 252 \text{ g}.$$

The 0 g in the "rate in" indicates there is no smoke entering into the room. The volume of 126 cubic meters is found by multiplying the room's dimensions, and the initial quantity of 252 g is found by multiplying the room's volume by 2. Simplified, we have

$$Q' = -\frac{1}{252}Q, \quad Q(0) = 252, \quad t \geq 0.$$

To solve we'll use the rule that when $y' = ky$, the solution is $y = Ce^{kt}$. Thus,

$$Q(t) = Ce^{(-1/252)t}.$$

The initial condition forces $C = 252$, so the specific solution is

$$Q(t) = 252e^{(-1/252)t}.$$

When does the room have 0.15 g of smoke per cubic meter? Remember, $Q(t)$ is the *total* quantity of smoke particles, so we really want to find t that makes $Q(t) = 0.15(126) = 18.9$ g, or

$$252e^{(-1/252)t} = 18.9$$

$$e^{(-1/252)t} = \frac{18.9}{252}$$

$$-\frac{1}{252}t = \ln\left(\frac{18.9}{252}\right)$$

$$t = -252\ln\left(\frac{18.9}{252}\right) \approx 652.7 \text{ minutes}.$$

Does the room ever get smoke-free? As $t \to \infty$, it is true $Q(t) \to 0$ as a limit. So yes, in a sense, the room becomes smoke free … as a limit.

Example 8.4: On August 1st, the pre-enrollment for Professor Jones' class "MAT900, Introduction to Hard Math", consists of 150 students of which 100 are seniors and 50 are juniors. Each day, 8 new students enroll, of which half are seniors and the other half are juniors. Also every day, 5 students drop the course, a mix of seniors and juniors. The first day of class is September 1st (31 days away). Approximately how many seniors will be enrolled on that day?

Solution: We start with the $\frac{dQ}{dt}$ = rate in − rate out form with reasonable adaptations. Note that the number of students increases by 3 per day.

$$\frac{dQ}{dt} = \left(\frac{4 \text{ seniors}}{8 \text{ students}}\right)\left(\frac{8 \text{ students}}{1 \text{ day}}\right) - \left(\frac{Q(t) \text{ seniors}}{(150 + 3t) \text{ students}}\right)\left(\frac{5 \text{ students}}{1 \text{ day}}\right).$$

Simplified, we have

$$Q' + \frac{5}{150 + 3t} Q = 4, \quad Q(0) = 100 \text{ seniors}, \quad 0 \leq t \leq 31.$$

Note that $t = 0$ represents August 1, so that $t = 30$ represents August 31, and $t = 31$ represents September 1.

The integration factor is

$$e^{\int (5/(150+3t))dt} = e^{(5/3)\ln(150+3t)} = (150 + 3t)^{5/3}.$$

The general solution is

$$Q(t) = \frac{\int 4(150 + 3t)^{5/3} dt + C}{(150 + 3t)^{5/3}}$$

$$= \frac{(1/2)(150 + 3t)^{8/3} + C}{(150 + 3t)^{5/3}}$$

$$= \frac{1}{2}(150 + 3t) + C(150 + 3t)^{-5/3}.$$

The initial condition is now incorporated:

$$100 = \frac{1}{2}(150 + 3(0)) + C(150 + 3(0))^{-5/3}$$

$$100 = 75 + C(150)^{-5/3}$$

$$25(150)^{5/3} = C$$

This C value can be left in its current form.

The specific solution is

$$Q(t) = \frac{1}{2}(150 + 3t) + 25(150)^{5/3}(150 + 3t)^{-5/3}.$$

When $t = 31$ we have

$$Q(31) = \frac{1}{2}(150 + 3(31)) + 25(150)^{5/3}(150 + 3(31))^{-5/3} \approx 132.687 \ldots$$

Thus, on September 1st, there should be about 133 seniors enrolled in the course.

As usual, we want to see if this is plausible. On August 1st, there were 100 seniors of out 150 students, a 2/3 (66.7%) ratio. Every day thereafter, only half (1/2, of 50%) of the incoming students were seniors. Thus, we would expect that the ratio of seniors to total students would decrease from 66.7%, but never get below 50%. As we see here, we expect that on the 31st say, we have 133 seniors out of $150 + 3(31) = 243$ students, a 133/243 ratio, or about 54.7%. This value agrees well with what we would expect, suggesting it is correct and that our analysis is solid.

See an error? Have a suggestion?
Please visit www.surgent.net/debook

Section 9
Numerical Methods: Euler and Heun Methods

There exist many numerical methods that allow us to construct an approximate solution to an ordinary differential equation. In this section, we will study two: Euler's Method, and Advanced Euler's (Heun's) Method.

Euler's Method:

Given: A differential equation of the form $\frac{dy}{dx} = f(x,y)$, with initial condition $y_0 = y(x_0)$.

Assume the solution exists over an interval $[x_0, b]$ and subdivide this interval into equal subdivisions of length h (the "step size"). Thus, a typical subinterval will have the form $[x_k, x_{k+1}]$, or equivalently, $[x_k, x_k + h]$.

Integrate both sides with respect to x from x_k to $x_k + h$:

$$\int_{x_k}^{x_k+h} \frac{dy}{dx} dx = \int_{x_k}^{x_k+h} f(x,y) \, dx.$$

The Fundamental Theorem of Calculus resolves the left side:

$$y(x_k + h) - y(x_k) = \int_{x_k}^{x_k+h} f(x,y) \, dx$$

$$y(x_k + h) = y(x_k) + \int_{x_k}^{x_k+h} f(x,y) \, dx.$$

Notation: We will call y_k the approximate value to $y(x_k)$, which represents an actual solution point of the differential equation.

We can approximate $\int_{x_k}^{x_k+h} f(x,y) \, dx$ using rectangles (similar to Riemann Sums). Thus, $\int_{x_k}^{x_k+h} f(x,y) \, dx \approx hf(x_k, y_k)$.

Thus, we have the following formula for approximating solutions to a differential equation:

$$y_{k+1} = y_k + hf(x_k, y_k).$$

This is a recursive formula. To find y_2 we have to know y_1 first, and so on.

Example 9.1: Find the approximate solutions of $\frac{dy}{dx} = x + y$ with $y(0) = 1$. Use a step size of $h = 0.1$.

Note: The initial condition is also written as $x_0 = 0$ and $y_0 = 1$. Also, $f(x, y)$ represents the right side of the differential equation, so $f(x, y) = x + y$. It does <u>not</u> represent the solution to the differential equation. That is $y = y(x)$, which is what we're trying to approximate.

Solution: The formula is $y_{k+1} = y_k + hf(x_k, y_k)$. Thus,

$$y_1 = y_0 + (0.1)f(x_0, y_0)$$
$$y_1 = 1 + (0.1)(0 + 1) \quad \text{remember, } f(x, y) = x + y$$
$$y_1 = 1 + 0.1$$
$$y_1 = 1.1.$$

We have a new approximation point, $(x_1, y_1) = (0.1, 1.1)$. Repeat the process:

$$y_2 = y_1 + 0.1f(x_1, y_1)$$
$$y_2 = 1.1 + 0.1(0.1 + 1.1)$$
$$y_2 = 1.1 + 0.1(1.2)$$
$$y_2 = 1.22.$$

We have another approximation point, $(x_2, y_2) = (0.2, 1.22)$.

$$y_3 = y_2 + 0.1f(x_2, y_2)$$
$$y_3 = 1.22 + 0.1(0.2 + 1.22)$$
$$y_3 = 1.362.$$

Now we have $(x_3, y_3) = (0.3, 1.362)$.

$$y_4 = y_3 + 0.1f(x_3, y_3)$$
$$y_4 = 1.362 + 0.1(0.3 + 1.362)$$
$$y_4 = 1.5282.$$

This gives us $(x_4, y_4) = (0.4, 1.5282)$. One more time:

$$y_5 = y_4 + 0.1f(x_4, y_4)$$
$$y_5 = 1.5282 + 0.1(0.4 + 1.5282)$$
$$y_5 = 1.72102.$$

This gives us $(x_5, y_5) = (0.5, 1.72102)$.

Let's compare these approximation points to those generated by the actual solution of $\frac{dy}{dx} = x + y$, $y(0) = 1$, which is found by using an integration factor. It is

$$y(x) = -x - 1 + 2e^x.$$

The following table shows the approximation points just generated, and the actual points provided by the solution:

x-value	Approximation	Actual
0	1	1
0.1	1.1	1.1103
0.2	1.22	1.2428...
0.3	1.362	1.3997...
0.4	1.5282	1.5836...
0.5	1.72102	1.7974...

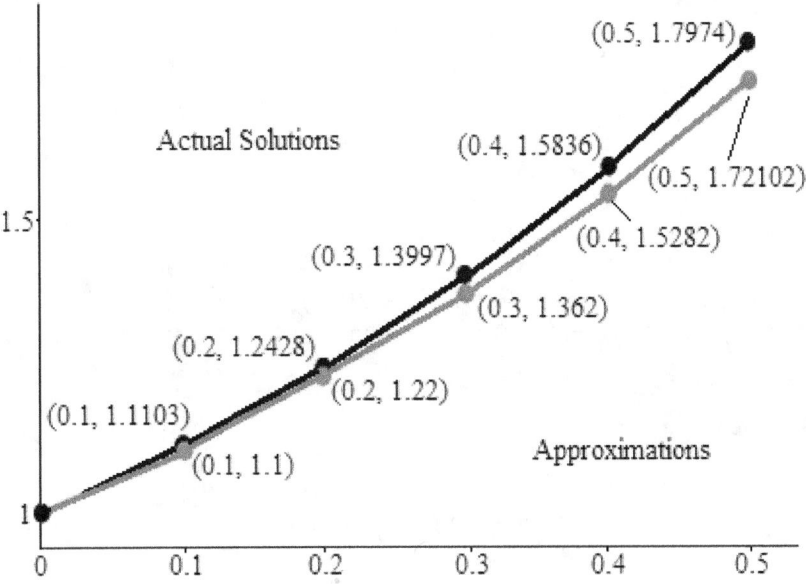

The graph of the actual solution (in black) in Example 9.1, and the approximations found by Euler's Method (in gray).

Example 9.2: Find the approximate solutions of $\frac{dy}{dx} = 1 - 2xy$ with $y(0) = 1$. Use a step size of $h = 0.1$.

Solution: This differential equation was given in Example 6.4. We attempted to use an integration factor to solve but could only go so far. We had:

$$y = \frac{\int e^{x^2} dx + C}{e^{x^2}}.$$

The integral $\int e^{x^2} dx$ is not renderable in common function form. We will generate a few approximation points instead of the solution.

Using the formula $y_{k+1} = y_k + hf(x_k, y_k)$, we have

$$y_1 = 1 + 0.1(1 - 2(0)(1))$$
$$y_1 = 1 + 0.1$$
$$y_1 = 1.1.$$

We have $(x_1, y_1) = (0.1, 1.1)$. Repeat:

$$y_2 = 1.1 + 0.1(1 - 2(0.1)(1.1))$$
$$y_2 = 1.1 + 0.1(0.78)$$
$$y_2 = 1.178.$$

We have $(x_2, y_2) = (0.2, 1.178)$. Repeat:

$$y_3 = 1.178 + 0.1(1 - 2(0.2)(1.178))$$
$$y_3 = 1.178 + 0.1(0.5288)$$
$$y_3 = 1.23088.$$

We have $(x_3, y_3) = (0.3, 1.23088)$.

$$y_4 = 1.23088 + 0.1(1 - 2(0.3)(1.23088))$$
$$y_4 = 1.23088 + 0.1(0.261472)$$
$$y_4 = 1.2570272.$$

We have $(x_4, y_4) = (0.4, 1.2570272)$. One last time:

$$y_5 = 1.2570272 + 0.1(1 - 2(0.4)(1.2570272))$$
$$y_5 = 1.2570272 + 0.1(0.44338944)$$
$$y_5 = 1.256465024.$$

We have $(x_5, y_5) = (0.5, 1.256465024)$.

Improved Euler's Method (also called Heun's Method):

Instead of using rectangles to approximate $\int_{x_k}^{x_k+h} f(x,y)\,dx$, we use trapezoids. A trapezoid with base h and heights $f(x_k, y_k)$ and $f(x_{k+1}, y_{k+1})$ has area

$$\frac{h}{2}[f(x_k, y_k) + f(x_{k+1}, y_{k+1})].$$

Thus, the formula now becomes

$$y_{k+1} = y_k + \frac{h}{2}[f(x_k, y_k) + f(x_{k+1}, y_{k+1})].$$

How do we approximate y_{k+1} on the left side when it's also part of the formula on the right side? It is replaced with the formula we used for Euler's Method:

$$y_{k+1} = y_k + \frac{h}{2}\left(f(x_k, y_k) + f\left(\underbrace{x_k + h}_{x_{k+1}}, \underbrace{y_k + hf(x_k, y_k)}_{y_{k+1}} \right) \right).$$

This is called the Improved Euler's Method, also known as Heun's Method.

Example 9.3: Use the Improved Euler's Method on $\frac{dy}{dx} = x + y$, $y(0) = 1$, with a step side of $h = 0.1$, to find the first five approximation points. (This is Example 9.1, and we will compare the results here with those derived using Euler's Method)

Solution: Using the formula, we have

$$y_1 = y_0 + \frac{0.1}{2}\left(f(x_0, y_0) + f(x_0 + 0.1, y_0 + 0.1 f(x_0, y_0)) \right)$$
$$y_1 = 1 + 0.05\left((0 + 1) + \left((0 + 0.1) + (1 + 0.1(0 + 1)) \right) \right)$$
$$y_1 = 1 + 0.05(1 + 0.1 + 1.1) = 1.11.$$

We have $(x_1, y_1) = (0.1, 1.11)$. Recall that Euler's Method gave $y_1 = 1.1$, and that the actual solution point is $y(0.1) = 1.1103$. Thus, we see that this method is already providing more precise approximations. Repeat:

$$y_2 = y_1 + 0.05\left((x_1 + y_1) + \left((x_1 + 0.1) + (y_1 + 0.1(x_1 + y_1)) \right) \right)$$
$$y_2 = 1.11 + 0.05\left((0.1 + 1.11) + \left((0.1 + 0.1) + (1.11 + 0.1(0.1 + 1.11)) \right) \right)$$
$$y_2 = 1.24205.$$

We have $(x_2, y_2) = (0.2, 1.24205)$. Repeat:

$$y_3 = 1.24205 + 0.05\Big((x_2 + y_2) + \big((x_2 + 0.1) + (y_2 + 0.1(x_2 + y_2))\big)\Big)$$
$$y_3 = 1.24205 + 0.05\Big((0.2 + 1.24205)$$
$$+ \big((0.2 + 0.1) + (1.24205 + 0.1(0.2 + 1.24205))\big)\Big)$$
$$y_3 = 1.39846\ldots$$

This gives $(x_3, y_3) = (0.3, 1.39846525)$.

$$y_4 = 1.39846\ldots + 0.05\Big((x_3 + y_3) + \big((x_3 + 0.1) + (y_3 + 0.1(x_3 + y_3))\big)\Big)$$
$$y_4 = 1.39846\ldots + 0.05\Big((0.3 + 1.39846\ldots)$$
$$+ \big((0.3 + 0.1) + (1.39846\ldots + 0.1(0.3 + 1.39846\ldots))\big)\Big)$$
$$y_4 = 1.581804\ldots$$

This gives $(x_4, y_4) = (0.4, 1.581804\ldots)$.

$$y_5 = 1.581804\ldots + 0.05\Big((x_3 + y_3) + \big((x_3 + 0.1) + (y_3 + 0.1(x_3 + y_3))\big)\Big)$$
$$y_5 = 1.581804\ldots + 0.05\Big((0.4 + 1.581804\ldots)$$
$$+ \big((0.4 + 0.1) + (1.581804\ldots + 0.1(0.4 + 1.581804\ldots))\big)\Big)$$
$$y_5 = 1.794893\ldots$$

This gives $(x_5, y_5) = (0.5, 1.794893\ldots)$.

Let's compare the two methods against the actual points:

x-value	Euler's	Heun's	Actual
0	1	1	1
0.1	1.1	1.11	1.1103
0.2	1.22	1.24205	1.2428…
0.3	1.362	1.3984…	1.3997…
0.4	1.5282	1.5818…	1.5836…
0.5	1.72102	1.7948…	1.7974…

The Improved Euler's/Heun's Method gives much better approximations, but at the cost of a lot more effort. These processes can be done on a spreadsheet, where one can vary the step size or run more rows as desired.

Compare and Contrast

These methods allow ways to find solution curves when the differential equation may not be solvable using analytical means. For example, it is impossible to find a "closed form" solution to $y' + 2xy = 1$. If we know an initial condition, we can numerically find approximate solutions to the differential equation.

The larger the step size, the approximations usually diverge faster from the actual solution. The smaller step sizes give better approximations but require more calculations to cover a certain interval.

Euler's method is fast but not as precise, while the Improved Euler's Method offers better precision, but takes more time.

Suggestion: do not round any calculations at any steps. This adds in "error", which is not desired since this is already an approximation technique. Write out all decimal places.

Write out each formula and each step since it's easy to get lost on each step.

Section 10
Higher-Order Linear Homogeneous & Autonomous Differential Equations with Constant Coefficients

In this presentation, we look at linear, nth-order autonomous and homogeneous differential equations with constant coefficients. Some examples are:

$$y'' + 4y' + 3 = 0$$
$$y^{(3)} - 9y' = 0$$
$$2y^{(4)} + 3y^{(3)} - y'' + y' + 6y = 0$$

We assume that a solution has the form $y = e^{rx}$, where r is a constant to be determined, since the expression e^{rx} will appear in its derivatives and can be factored to the front.

Example 10.1: Find the general solution of $y'' - 7y' + 12y = 0$.

Solution: Let $y = e^{rx}$. Therefore, $y' = re^{rx}$ and $y'' = r^2 e^{rx}$. Substituting, we have

$$(r^2 e^{rx}) - 7(re^{rx}) + 12e^{rx} = 0$$
$$e^{rx}(r^2 - 7r + 12) = 0$$
$$e^{rx}(r - 3)(r - 4) = 0$$
$$r = 3, \ r = 4.$$

Thus, possible solutions are $y = e^{3x}$ and $y = e^{4x}$. These are easy to check: for $y = e^{3x}$, we have $y' = 3e^{3x}$ and $y'' = 9e^{3x}$. Substituting, we have

$$(9e^{3x}) - 7(3e^{3x}) + 12(e^{3x}) = e^{3x}(9 - 21 + 12) = e^{3x}(0) = 0$$

For $y = e^{4x}$, we have $y' = 4e^{4x}$ and $y'' = 16e^{4x}$. Substituting, we have

$$(16e^{3x}) - 7(4e^{3x}) + 12(e^{3x}) = e^{3x}(16 - 28 + 12) = e^{3x}(0) = 0.$$

The **Law of Superposition** states that if y_1 and y_2 are linearly independent solutions of a differential equation of the form we are discussing, then so is their linear product: $y = C_1 y_1 + C_2 y_2$. In our example, the general solution is $y = C_1 e^{3x} + C_2 e^{4x}$. (We'll discuss "linear independence" momentarily.)

We check it: $y' = 3C_1 e^{3x} + 4C_2 e^{4x}$ and $y'' = 9C_1 e^{3x} + 16C_2 e^{4x}$. Substitute:

$$\overbrace{(9C_1 e^{3x} + 16C_2 e^{4x})}^{y''} - 7\overbrace{(3C_1 e^{3x} + 4C_2 e^{4x})}^{y'} + 12\overbrace{(C_1 e^{3x} + C_2 e^{4x})}^{y} = 0$$
$$9C_1 e^{3x} - 21C_1 e^{3x} + 12C_1 e^{3x} + 16C_2 e^{4x} - 28C_2 e^{4x} + 12C_2 e^{4x} = 0$$
$$C_1 e^{3x}(9 - 21 + 12) + C_2 e^{4x}(16 - 28 + 12) = 0$$
$$C_1 e^{3x}(0) + C_2 e^{4x}(0) = 0.$$

The domain is all real numbers. This demonstrates that the linear product of the individual solutions forms a general solution.

In the above example, the equation $r^2 - 7r + 12 = 0$ is called the **auxiliary polynomial**. Normally, we can move directly from the given differential equation to the auxiliary polynomial and skip the handful of steps in between.

Example 10.2: Find the general solution of $y'' - 4y' - 12y = 0$.

Solution: The auxiliary polynomial set to zero is $r^2 - 4r - 12 = 0$. Factored, this is $(r - 6)(r + 2) = 0$, so that the solutions are $r = 6$ and $r = -2$. Therefore, the general solution is

$$y = C_1 e^{6x} + C_2 e^{-2x}.$$

Example 10.3: Find the general solution of $y^{(3)} - 9y' = 0$.

Solution: The auxiliary polynomial set to zero is $r^2 - 9r = 0$. Factored, we have $r(r + 3)(r - 3) = 0$. The solutions are $r = 0, r = 3$ and $r = -3$, and the general solution is

$$y = C_1 + C_2 e^{3x} + C_3 e^{-3x}.$$

In this case, we simplified $C_1 e^{0x} = C_1(1) = C_1$.

Example 10.4: Find the general solution of $y'' - 4y' + 1 = 0$.

Solution: The auxiliary polynomial set to zero is $r^2 - 4r + 1 = 0$. The quadratic formula is used to determine the roots:

$$r = \frac{-(-4) \pm \sqrt{(-4)^2 - 4(1)(1)}}{2(1)} = \frac{4 \pm \sqrt{12}}{2} = \frac{4 \pm 2\sqrt{3}}{2} = 2 \pm \sqrt{3}.$$

The general solution is

$$y = C_1 e^{(2+\sqrt{3})x} + C_2 e^{(2-\sqrt{3})x}.$$

A second-order linear homogeneous ordinary differential equation will have a general solution of the form $y = C_1 y_1 + C_2 y_2$, where y_1 and y_2 are the individual component solutions such as those found in the previous examples. They form a *solution basis*, in which all possible solutions are some linear combination of the component solutions. Linear independence is discussed more in depth in the following section.

Example 10.5: Find the general solution of $y^{(3)} + y'' - 4y' - 4y = 0$.

Solution: The auxiliary polynomial set to zero is $r^3 + r^2 - 4r - 4 = 0$. This is a cubic polynomial with no easy way to factor it. Instead, we locate the roots by viewing its graph and noting where it crosses the horizontal axis:

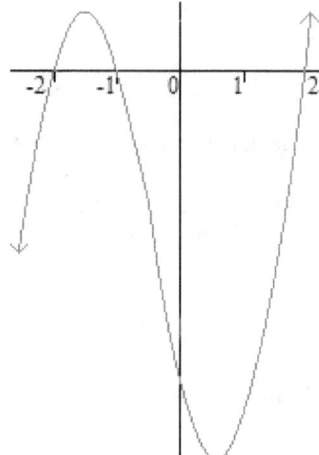

The graph of $y = r^3 + r^2 - 4r - 4$ has roots $r = -2, r = -1$ and $r = 2$.

Thus, the general solution is

$$y = C_1 e^{-2x} + C_2 e^{-x} + C_3 e^{2x}.$$

Example 10.6: Find the specific solution of the IVP $y'' - 2y' - 15y = 0$ with initial conditions $y(0) = 9, y'(0) = 29$.

Solution: The auxiliary polynomial set to zero is $r^2 - 2r - 15 = 0$. It factors as $(r - 5)(r + 3) = 0$, giving two roots, $r = 5$ and $r = -3$. Thus, the general solution is

$$y = C_1 e^{5x} + C_2 e^{-3x}.$$

To find C_1 and C_2, we need the first derivative of the general solution, which is

$$y' = 5C_1 e^{5x} - 3C_2 e^{-3x}.$$

Now the initial conditions are considered. When $y(0) = 9$, we have

$$C_1 e^{5(0)} + C_2 e^{-3(0)} = 9$$
$$C_1 + C_2 = 9.$$

When $y'(0) = 29$, we have

$$5C_1 e^{5(0)} - 3C_2 e^{-3(0)} = 29$$
$$5C_1 - 3C_2 = 29.$$

This is a linear system of two equations and two variables:

$$C_1 + C_2 = 9$$
$$5C_1 - 3C_2 = 29.$$

To solve, there are options. In this case, we can multiply one row by a constant then add, in such a way that one of the C values momentarily drops out. Let's multiply the top equation by 3:

$$3C_1 + 3C_2 = 27$$
$$5C_1 - 3C_2 = 29.$$

Summing the two equations, C_2 momentarily drops out, and we have

$$8C_1 = 56.$$

Thus, $C_1 = 7$. Now, substitute this value into one of the equations to find C_2. We get $C_2 = 2$.

Thus, the solution of the IVP is

$$y = 7e^{5x} + 2e^{-3x}.$$

Note: for an nth ordered differential equation, n initial conditions will be necessary to solve for the leading constants.

Thus far in this section, we have only looked at differential equations with distinct real roots (that is, each has multiplicity 1). Momentarily, we will look at cases where one root may have multiplicity greater than 1, and cases where the roots are complex.

But first, we discuss the notion of linear independence more in depth. This is critical to understand how general solutions are composed.

Section 11
Linear Independence of Solutions: The Wronskian

A **linear combination** of two functions (terms, etc.) $f(x)$ and $g(x)$ is a single expression in the form

$$af(x) + bg(x),$$

where a and b are constants called scalars. (This can be applied for 3 or more terms too).

For example, the linear combination of x^2 and x is

$$ax^2 + bx.$$

We assume, for the moment, that these terms cannot be combined as like terms.

Two equations (functions, component solutions, etc.) are **linearly independent** is one is not a non-zero scalar multiple of the other.

For example, $f(x) = 2x$ and $g(x) = 3x^2$ are linearly independent because one cannot be written as a scalar multiple of the other. However, $f(x) = 2x$ and $g(x) = 4x$ are not linearly independent, since $g(x) = 2f(x)$.

In a linear system of two equations and two variables, the equations are linearly independent if one row is not a scalar multiple of the other row. For example, the system

$$3x - y = 8$$
$$-x + 2y = 11$$

is linearly independent since neither row is a scalar multiple of the other. However, the system

$$5x + 2y = 3$$
$$15x + 6y = 9$$

is not linearly independent since the bottom row is 3 times the top row, or equivalently, the top row is 1/3 times the bottom row.

This is important because the usual method making one of the unknowns "drop out" works if the equations are linearly independent and fails if the equations are not. Try solving for x or y in the preceding system and you'll see why it will fail.

For larger systems, it is more difficult to determine quickly if the equations are linearly independent. For larger systems, if any one row is equal to some linear combination of two or more rows in the system, then the rows are not linearly independent. However, this is difficult to see by observation alone. We need a more powerful method.

Linear Independence and the Wronskian

Let $y = C_1 f(x) + C_2 g(x)$ be the general solution of a linear second-order homogeneous differential equation, and assume it has initial conditions $y(x_0) = A$ and $y'(x_0) = B$, where A and B are any two real numbers.

To find A and B, we need the derivative: $y' = C_1 f'(x_0) + C_2 g'(x_0)$. Now the initial conditions are considered:

$$\begin{matrix} y(x_0) = A \\ y'(x_0) = B \end{matrix} \text{ gives } \begin{matrix} A = C_1 f(x_0) + C_2 g(x_0) \\ B = C_1 f'(x_0) + C_2 g'(x_0) \end{matrix}$$

We form a 2 × 2 matrix of the functions and their derivatives:

$$\begin{bmatrix} f(x_0) & g(x_0) \\ f'(x_0) & g'(x_0) \end{bmatrix}.$$

The determinant of a 2 × 2 matrix is the product of the main diagonal minus the product of the secondary diagonal:

$$\det \begin{bmatrix} f(x_0) & g(x_0) \\ f'(x_0) & g'(x_0) \end{bmatrix} = f(x_0) g'(x_0) - f'(x_0) g(x_0).$$

For a solution to exist, the determinant must not be 0. (This and other aspects of matrices are discussed later in Section 25).

This is called the Wronskian. Its notation is

$$W(f, g) = \det \begin{bmatrix} f(x_0) & g(x_0) \\ f'(x_0) & g'(x_0) \end{bmatrix}.$$

We use the Wronskian to verify if the individual solutions of a differential equation are linearly independent of one another, or not. If not, simply combine them as like terms. But in doing so, other solutions may be ignored.

Example 11.1: Show that the expressions e^{6x} and e^{-2x} are linearly independent (these are the individual solutions from Example 10.2).

Solution: We find the Wronskian:

$$W(e^{6x}, e^{-2x}) = \det \begin{bmatrix} e^{6x} & e^{-2x} \\ 6e^{6x} & -2e^{-2x} \end{bmatrix}$$

$$= (e^{6x})(-2e^{-2x}) - (e^{-2x})(6e^{6x})$$

$$= -2e^{4x} - 6e^{4x}$$

$$= -8e^{4x}.$$

The expression $-8e^{4x}$ is never 0 for all real numbers x. Thus, $W(e^{6x}, e^{-2x}) \neq 0$ and the expressions are linearly independent. It is "okay" to combine them in a linear combination to form the general solution:

$$y = C_1 e^{4x} + C_2 e^{-3x}.$$

In general, the following forms are linearly independent:

Exponentials e^{ax} and e^{bx} are linearly independent when $a \neq b$. This can be extended to three or more such expressions.

Polynomial terms such as x^2, $3x$ and 2 are linearly independent. In simpler terms, if they were not linearly independent, we would combine them as like terms.

Trigonometric terms such as $\sin ax$ and $\sin bx$ are linearly independent when $a \neq b$. Expressions such as $\cos x$ and $\sin x$ are linearly independent.

Expressions that are structurally different, such as e^{5t} and te^{5t} will be linearly independent.

If in doubt, find the Wronskian.

The determinant of a 3×3 matrix is given by the formula

$$\det \begin{bmatrix} a_1 & a_2 & a_3 \\ b_1 & b_2 & b_3 \\ c_1 & c_2 & c_3 \end{bmatrix} = a_1 \det \begin{bmatrix} b_2 & b_3 \\ c_2 & c_3 \end{bmatrix} - a_2 \det \begin{bmatrix} b_1 & b_3 \\ c_1 & c_3 \end{bmatrix} + a_3 \det \begin{bmatrix} b_1 & b_2 \\ c_1 & c_2 \end{bmatrix}.$$

Example 11.2: Consider these three functions:

$$y_1(x) = x^2 + 2x, \quad y_2(x) = 3x + 1, \quad y_3(x) = 2x^2 + x - 1.$$

Determine whether they are linearly independent or not.

Solution: We have

$$W(y_1, y_2, y_3) = \det \begin{bmatrix} x^2 + 2x & 3x + 1 & 2x^2 + x - 1 \\ 2x + 2 & 3 & 4x + 1 \\ 2 & 0 & 4 \end{bmatrix}$$

$$= (x^2 + 2x) \det \begin{bmatrix} 3 & 4x + 1 \\ 0 & 4 \end{bmatrix}$$

$$- (3x + 1) \det \begin{bmatrix} 2x + 2 & 4x + 1 \\ 2 & 4 \end{bmatrix}$$

$$+ (2x^2 + x - 1) \det \begin{bmatrix} 2x + 2 & 3 \\ 2 & 0 \end{bmatrix}$$

$$= (x^2 + 2x)(12) - (3x + 1)(6) + (2x^2 + x - 1)(-6)$$

$$= 12x^2 + 24x - 18x - 6 - 12x^2 - 6x + 6$$

$$= 12x^2 - 12x^2 + 24x - 18x - 6x + 6 - 6 = 0.$$

These equations are *not* linearly independent. Note that $2y_1(x) - y_2(x) = y_3(x)$. This means (by example) that each equation is some linear combination of the other two. But would you have known it just by looking?

Section 12
Complex Roots of the Auxiliary Polynomial

We start with an example:

Example 12.1: Find the general solution of $y'' + y = 0$.

Solution: From Example 2.6, we know that $y_1 = \cos x$ and $y_2 = \sin x$ both solve this differential equation. Thus, the general solution is

$$y = C_1 \cos x + C_2 \sin x.$$

Alternatively, assume the solution has the e^{rx} form, the auxiliary polynomial set to zero is $r^2 + 1 = 0$, which has roots $\pm i$. Thus, the general solution is

$$y = C_1 e^{ix} + C_2 e^{-ix}.$$

Two different methods give two different solutions. Or are they different?

Review of Maclaurin Series

The Maclaurin Series for e^x, $\cos x$ and $\sin x$ are given below. Each is centered at $x = 0$ and each has a domain of $-\infty < x < \infty$.

$$e^x = 1 + x + \frac{1}{2!}x^2 + \frac{1}{3!}x^3 + \frac{1}{4!}x^4 + \frac{1}{5!}x^5 + \frac{1}{6!}x^6 + \frac{1}{7!}x^7 + \cdots,$$

$$\cos x = 1 - \frac{1}{2!}x^2 + \frac{1}{4!}x^4 - \frac{1}{6!}x^6 + \frac{1}{8!}x^8 - \frac{1}{10!}x^{10} + \frac{1}{12!}x^{12} - \cdots,$$

$$\sin x = x - \frac{1}{3!}x^3 + \frac{1}{5!}x^5 - \frac{1}{7!}x^7 + \frac{1}{9!}x^9 - \frac{1}{11!}x^{11} + \frac{1}{13!}x^{13} - \cdots.$$

Using the Maclaurin Series for e^x, replace x with ix:

$$e^{ix} = 1 + ix + \frac{1}{2!}(ix)^2 + \frac{1}{3!}(ix)^3 + \frac{1}{4!}(ix)^4 + \frac{1}{5!}(ix)^5 + \frac{1}{6!}(ix)^6 + \frac{1}{7!}(ix)^7 + \cdots.$$

The powers of the imaginary unit i are: $i^2 = -1$, $i^3 = -i$, $i^4 = 1$, $i^5 = i$, and so on, repeating the pattern. Thus, the above line is now:

$$e^{ix} = 1 + ix - \frac{1}{2!}x^2 - \frac{1}{3!}ix^3 + \frac{1}{4!}x^4 + \frac{1}{5!}ix^5 - \frac{1}{6!}x^6 - \frac{1}{7!}ix^7 + \frac{1}{8!}x^8 + \cdots.$$

The even-powered terms no longer contain i, while the odd-power terms do.

Now, regroup the terms and simplify:

$$e^{ix} = \underbrace{\left(1 - \frac{1}{2!}x^2 + \frac{1}{4!}x^4 - \frac{1}{6!}x^6 + \frac{1}{8!}x^8 - \cdots\right)}_{\cos x} + i\underbrace{\left(x - \frac{1}{3!}x^3 + \frac{1}{5!}x^5 - \frac{1}{7!}x^7 + \frac{1}{9!}x^9 - \cdots\right)}_{\sin x}$$

This gives the famous identity known as Euler's Formula:

$$e^{ix} = \cos x + i \sin x.$$

Note that $e^{-ix} = \cos x - i \sin x$. Thus, making the substitutions,

$$y = C_1 e^{ix} + C_2 e^{-ix}$$

becomes

$$y = C_1(\cos x + i \sin x) + C_2(\cos x - i \sin x).$$

This simplifies to

$$y = (C_1 + C_2) \cos x + i(C_1 - C_2) \sin x.$$

Since C_1 and C_2 have no meaning yet, we can replace $C_1 + C_2$ with "new" C_1 and $C_1 - C_2$ with "new" C_2. Now we have

$$y = C_1 \cos x + iC_2 \sin x.$$

Can we remove the imaginary unit i somehow? We can, by the following argument:

If $y = f(x) + ig(x)$ is a solution to an ordinary homogeneous differential equation with constant coefficients, then $f(x)$ and $g(x)$ can be treated as individual solutions.

Assume that $y = f(x) + ig(x)$ is a solution to $Ay'' + By' + Cy = 0$. Differentiating, we have $y' = f'(x) + ig'(x)$ and $y'' = f''(x) + ig''(x)$. Substituting we have

$$A\bigl(f''(x) + ig''(x)\bigr) + B\bigl(f'(x) + ig'(x)\bigr) + C\bigl(f(x) + ig(x)\bigr) = 0$$
$$\bigl(Af''(x) + Bf'(x) + Cf(x)\bigr) + i\bigl(Ag''(x) + Bg'(x) + Cg(x)\bigr) = 0.$$

For this to work, we treat 0 as a complex number, $0 + i0$, so that we have

$$\bigl(Af''(x) + Bf'(x) + Cf(x)\bigr) + i\bigl(Ag''(x) + Bg'(x) + Cg(x)\bigr) = 0 + i0.$$

This implies that $Af''(x) + Bf'(x) + Cf(x) = 0$, so that $f(x)$ is a solution alone, and that $i\bigl(Ag''(x) + Bg'(x) + Cg(x)\bigr) = i0$, which is the same as $Ag''(x) + Bg'(x) + Cg(x) = 0$, so that $g(x)$ is also a solution. When f and g are linearly independent, they form a general solution.

General rule: If an ODE is of second order, homogeneous and with constant coefficients, and its auxiliary polynomial has complex roots $x = a \pm bi$, then the general solution is

$$y = C_1 e^{ax} \cos bx + C_2 e^{ax} \sin bx.$$

Example 12.2: Find the general solution of $y'' + 2y' + 5y = 0$.

Solution: The auxiliary polynomial is $r^2 + 2r + 5 = 0$. Using the quadratic formula, the roots are $y = -1 \pm 2i$, which gives $a = -1$ and $b = 2$. Thus, the general solution is

$$y = C_1 e^{-x} \cos 2x + C_2 e^{-x} \sin 2x.$$

Are the component functions of the solution linearly independent? The derivatives are:

$$y_1' = e^{-x}(\cos 2x - 2\sin 2x) \text{ and } y_2' = e^{-x}(2\cos 2x - \sin 2x)$$

Thus, we have

$$W(y_1, y_2) = \det \begin{bmatrix} e^{-x} \cos 2x & e^{-x} \sin 2x \\ e^{-x}(\cos 2x - 2\sin 2x) & e^{-x}(2\cos 2x - \sin 2x) \end{bmatrix}$$

$$= e^{-2x}(2\cos^2 2x - \cos 2x \sin 2x) - e^{-2x}(-2\sin^2 2x - \cos 2x \sin 2x)$$

$$= e^{-2x}(\underbrace{2\cos^2 2x + 2\sin^2 2x}_{2} \underbrace{- \cos 2x \sin 2x + \cos 2x \sin 2x}_{0})$$

$$= 2e^{-2x}.$$

Since $2e^{-2x}$ is never 0, the component functions $e^{-x} \cos 2x$ and $e^{-x} \sin 2x$ are linearly independent, and that

$$y = C_1 e^{-x} \cos 2x + C_2 e^{-x} \sin 2x$$

is the general solution of $y'' + 2y' + 5y = 0$. In general, $e^{ax} \cos bx$ and $e^{ax} \sin bx$ are always linearly independent.

Example 12.3: Find the specific solution to the IVP $y^{(4)} - y = 0$, with initial conditions $y(0) = 1$, $y'(0) = 2$, $y''(0) = -1$, and $y^{(3)}(0) = 1$.

Solution: The auxiliary polynomial is $r^4 - 1 = 0$, which factors as a difference of squares into $(r^2 - 1)(r^2 + 1)$ and then again to $(r+1)(r-1)(r^2+1)$.

From $(r+1)$, we get a root $r = -1$, so that $y = e^{-x}$ is a solution.

From $(r-1)$, we get a root $r = 1$, so that $y = e^x$ is a solution.

From $(r^2 + 1)$, we get the complex solutions $y = \pm i$, or $0 \pm 1i$, so that $a = 0$ and $b = 1$. Thus, we get $y = e^{0x} \cos(1x) + e^{0x} \sin(1x)$ as solutions, which can be simplified to $y = \cos x + \sin x$. Thus, the general solution is

$$y = C_1 e^{-x} + C_2 e^x + C_3 \cos x + C_4 \sin x.$$

To find the specific solution, we need to differentiate three times:

$$y' = -C_1 e^{-x} + C_2 e^x - C_3 \sin x + C_4 \cos x,$$
$$y'' = C_1 e^{-x} + C_2 e^x - C_3 \cos x - C_4 \sin x,$$
$$y^{(3)} = -C_1 e^{-x} + C_2 e^x + C_3 \sin x - C_4 \cos x.$$

The initial conditions are now considered:

$y(0) = 1$ gives $C_1 e^{-0} + C_2 e^0 + C_3 \cos 0 + C_4 \sin 0 = 1$,
$y'(0) = 2$ gives $-C_1 e^{-0} + C_2 e^0 - C_3 \sin 0 + C_4 \cos 0 = 2$,
$y''(0) = -1$ gives $C_1 e^{-0} + C_2 e^0 - C_3 \cos 0 - C_4 \sin 0 = -1$,
$y^{(3)}(0) = 1$ gives $-C_1 e^{-0} + C_2 e^0 + C_3 \sin 0 - C_4 \cos 0 = 1$.

Recall that $e^0 = 1, \cos 0 = 1$ and $\sin 0 = 0$. Thus, the above set of equations becomes a four-equation, linear system where the four constants are the unknowns:

$$C_1 + C_2 + C_3 = 1$$
$$-C_1 + C_2 + C_4 = 2$$
$$C_1 + C_2 - C_3 = -1$$
$$-C_1 + C_2 - C_4 = 1.$$

To solve systems of this size is usually better left to technology. Most calculators have built-in features that will simplify a system and provide the solution in seconds (see page 175 for a discussion). The constants are $C_1 = -\frac{3}{4}$, $C_2 = \frac{3}{4}$, $C_3 = 1$ and $C_4 = \frac{1}{2}$. Thus, the solution to the IVP is

$$y = -\frac{3}{4} e^{-x} + \frac{3}{4} e^x + \cos x + \frac{1}{2} \sin x.$$

The domain is all real values x. The four basis solutions are linearly independent of one another.

The cosine-combined form of a solution:
Converting from the $C_1 \cos bt + C_2 \sin bt$ form to the $A \cos(bt - P)$ form.

Suppose a solution of an IVP is

$$y = 0.2 \cos 3t + 0.3 \sin 3t.$$

This can be rewritten in the form $y = A \cos(3t - P)$, where A is the amplitude and P is the phase shift. This is called the *cosine-combined form* and has the advantage of being described by one trigonometric function, not two, and that its amplitude and any horizontal or phase shifts are easily gleaned from the equation.

We start with the cosine-difference identity:

$$\cos(\alpha - \beta) = \cos \alpha \cos \beta + \sin \alpha \sin \beta.$$

Let $\alpha = P$ and $\beta = 3t$, and multiply through by A:

$$A \cos(P - 3t) = A \cos P \cos 3t + A \sin P \sin 3t.$$

Note that $\cos(3t - P) = \cos(-P + 3t) = \cos(P - 3t)$ using the symmetry rule that $\cos(-t) = \cos t$.

We have

$$A \cos(P - 3t) = \underbrace{A \cos P}_{0.2} \cos 3t + \underbrace{A \sin P}_{0.3} \sin 3t.$$

This makes $A \cos P = 0.2$ and $A \sin P = 0.3$. Thus, $\cos P = \frac{0.2}{A}$ and $\sin P = \frac{0.3}{A}$. Since $\cos^2 P + \sin^2 P = 1$

$$\left(\frac{0.2}{A}\right)^2 + \left(\frac{0.3}{A}\right)^2 = 1$$

$$(0.2)^2 + (0.3)^2 = A^2$$

$$A = \sqrt{0.04 + 0.09} = \sqrt{0.13}.$$

Since $\tan P = \frac{\sin P}{\cos P}$, we have

$$\tan P = \frac{0.3/A}{0.2/A} = \frac{0.3}{0.2}.$$

Thus, $P = \tan^{-1}\left(\frac{0.3}{0.2}\right) \approx 0.983$ radians.

Assembling this together, we have

$$y = 0.2 \cos 3t + 0.3 \sin 3t = \sqrt{0.13} \cos(3t - 0.983).$$

In general, if given $C_1 \cos bt + C_2 \sin bt$, then

$$A = \sqrt{C_1^2 + C_2^2} \quad \text{and} \quad P = \tan^{-1}\left(\frac{C_2}{C_1}\right).$$

Be careful when assembling the quotient for the arctangent calculation. All angle figures are in radians.

From the cosine-combined form, $y = \sqrt{13} \cos(3t - 0.983)$, we can rewrite it slightly differently, factoring the 3 inside the cosine operator:

$$\sqrt{13} \cos(3(t - 0.328)).$$

The amplitude is $A = \sqrt{0.13} \approx 0.361$ units, the phase shift is 0.983 radians, but the actual horizontal shift is 0.328 radians.

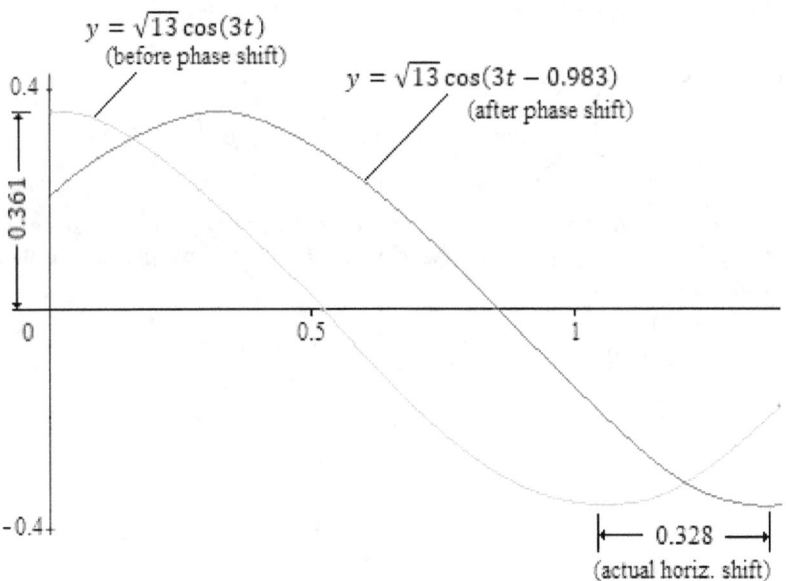

Section 13
Repeated Roots of the Auxiliary Polynomial

Consider the second-order differential equation $y'' - 6y' + 9y = 0$.

Its auxiliary polynomial equation is $r^2 - 6r + 9 = 0$, which factors as $(r-3)(r-3) = 0$. Thus, $r = 3$ is a root with multiplicity 2.

One solution is $y_1 = C_1 e^{3x}$. However, since 3 is a root with multiplicity 2, we cannot write the other solution as $y_2 = C_2 e^{3x}$. In other words, the general solution is not $y = C_1 e^{3x} + C_2 e^{3x}$. The two presumptive component functions are not linearly independent.

There does exist another solution to this differential equation, one that is linearly independent of $y_1 = e^{3x}$. It is $y_2 = xe^{3x}$. To verify this is true, note that the derivatives are $y_2' = 3xe^{3x} + e^{3x}$ and $y_2'' = 9xe^{3x} + 6e^{3x}$. Substitute:

$$(9xe^{3x} + 6e^{3x}) - 6(3xe^{3x} + e^{3x}) + 9(xe^{3x}) = 0.$$

Distribute to clear parentheses:

$$9xe^{3x} + 6e^{3x} - 18xe^{ex} - 6e^{3x} + 9xe^{3x} = 0.$$

Collect like terms and add:

$$\underbrace{(9xe^{3x} - 18xe^{3x} + 9xe^{3x})}_{0} + \underbrace{(6e^{3x} - 6e^{3x})}_{0} = 0.$$

The component functions e^{3x} and xe^{3x} are not linearly dependent since it is impossible to multiply one by a constant to get the other. Nevertheless, we check it with the Wronskian:

$$W(e^{3x}, xe^{3x}) = \det \begin{bmatrix} e^{3x} & xe^{3x} \\ 3e^{3x} & 3xe^{3x} + e^{3x} \end{bmatrix}$$

$$= (e^{3x})(3xe^{3x} + e^{3x}) - (3e^{3x})(xe^{3x})$$

$$= 3xe^{6x} + e^{6x} - 3xe^{6x}$$

$$= e^{6x}.$$

Since e^{6x} is never 0 for all real-values x in the domain, the two component functions are linearly independent and form the basis of the general solution:

$$y = C_1 e^{3x} + C_2 xe^{3x}.$$

The justification for the leading x in the second term is given in subsequent Section 14, Reduction of Order.

General Rule: If r is a real-valued root with multiplicity n of the auxiliary polynomial of a linear, homogenous ODE with constant coefficients, then it provides n linearly independent solutions, which are

$$y_1 = e^{rx}, \; y_2 = xe^{rx}, \; y_3 = x^2 e^{rx}, \; \ldots, \; y_n = x^{n-1} e^{rx}.$$

Example 13.1: Find the general solution of $y^{(3)} + 3y'' + 3y' + y = 0$.

Solution: The auxiliary polynomial is $r^3 + 3r^2 + 3r + 1 = 0$, which factors as $(r+1)^3 = 0$. Thus, $r = -1$ is the root of this polynomial, multiplicity 3.

Using the above rule, the individual component solutions are

$$y_1 = e^{-x}, \; y_2 = xe^{-x} \text{ and } y_3 = x^2 e^{-x},$$

And the general solution is

$$y = C_1 e^{-x} + C_2 x e^{-x} + C_3 x^2 e^{-x}.$$

Repeated Complex Roots

General Rule: If $r = a \pm bi$ is a pair of conjugate complex-valued roots with multiplicity n each of the auxiliary polynomial of a linear, homogenous ODE with constant coefficients, then it provides $2n$ linearly independent solutions, which are

$$y_{1,2} = C_1 e^{ax} \cos bx + C_2 e^{ax} \sin bx,$$
$$y_{3,4} = C_3 x e^{ax} \cos bx + C_4 x e^{ax} \sin bx, \text{ and so on.}$$

Example 13.2: Find the general solution of $y^{(4)} + 8y'' + 16y = 0$.

Solution: The auxiliary polynomial is $r^4 + 8r^2 + 16 = 0$, which factors as $(r^2 + 4)^2 = 0$. Thus, $r = \pm 2i$ are roots, each of multiplicity 2. The basis solutions are

$$y_1 = \cos 2x, \; y_2 = \sin 2x, \; y_3 = x \cos 2x, \; y_4 = x \sin 2x,$$

And the general solution is

$$y = C_1 \cos 2x + C_2 \sin 2x + C_3 x \cos 2x + C_4 x \sin 2x.$$

Example 13.3: Find the general solution of

$$y^{(4)} + 4y^{(3)} + 14y'' + 20y' + 25y = 0.$$

Solution: The auxiliary polynomial is $r^4 + 4r^3 + 14r^2 + 20r + 25 = 0$. This factors as $(r^2 + 2r + 5)^2 = 0$. The roots are $r = -1 \pm 2i$, each multiplicity 2.

The general solution is

$$y = C_1 e^{-x} \cos 2x + C_2 e^{-x} \sin 2x + C_3 x e^{-x} \cos 2x + C_4 x e^{-x} \sin 2x.$$

There is no general way to factor quartic polynomials. The above polynomial was factored using online computer-algebra websites.

Graphical methods may help locate roots of a higher-degree polynomial.

Example 13.4: Find the general solution of

$$y^{(6)} - 9y^{(5)} + 12y^{(4)} + 76y^{(3)} - 144y'' - 192y' + 256 = 0.$$

Solution: the auxiliary polynomial is

$$r^6 - 9r^5 + 12r^4 + 76r^3 - 144r^2 - 192r + 256 = 0.$$

Factoring this sixth-degree polynomial is difficult. Instead, we inspect its graph and study where it meets the horizontal axis, and how it crosses or touches it:

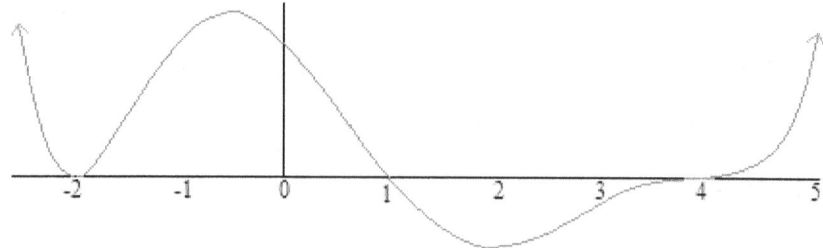

- If the graph passes directly through the horizontal axis without pausing, it is a root of multiplicity 1. Thus, in this case, $r = 1$, multiplicity 1.

- If the graph touches the horizontal axis tangentially but never crosses it, it is likely a root of multiplicity 2. In the case, $r = -2$, multiplicity 2.

- If the graph passes through the horizontal axis but pauses as it does so, in the sense that the horizontal axis is tangential to the graph at this point, then it is likely a root of multiplicity 3. In this case, $r = 4$, multiplicity 3.

This suggests that the auxiliary polynomial factors as

$$(r-1)(r+2)^2(r-4)^3.$$

Note that the multiplicities add to 6, the degree of the polynomial. This means we have not ignored any other possible roots.

The general solution is

$$y = C_1 e^x + C_2 e^{-2x} + C_3 x e^{-2x} + C_4 e^{4x} + C_5 x e^{4x} + C_6 x^2 e^{4x}.$$

By now, we have already established that e^{ax} and e^{bx} are linearly independent if $a \neq b$, and that e^{ax}, xe^{ax}, $x^2 e^{ax}$, and so on, are linearly independent. Thus, these six component functions are linearly independent, and performing the determinant of a 6×6 Wronskian matrix is not necessary.

Example 13.5: Find the general solution of

$$y^{(5)} - 2y^{(4)} - 5y^{(3)} - 2y'' + 52y' - 56y = 0.$$

Solution: The auxiliary polynomial is $r^5 - 2r^4 - 5r^3 - 2r^2 + 52r - 56$. Its graph is shown below.

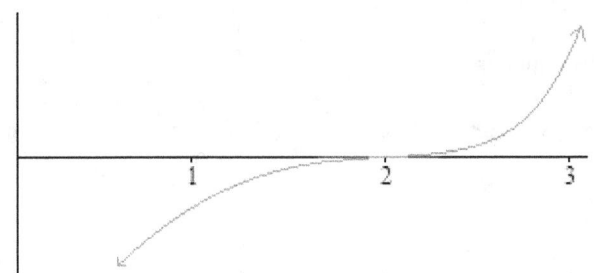

This is a close-up of the only place where the graph crosses the horizontal axis. There appears to be a root at $r = 2$, and that it passes tangentially through the horizontal axis, so it probably has multiplicity 3. However, we need to verify this using polynomial division. Synthetic division is a fast way to perform polynomial division, one linear factor at a time. We are dividing by $r - 2$:

```
2 | 1  -2  -5  -2   52  -56
  |      2   0  -10  -24   56
  ----------------------------
    1   0  -5  -12  28  | 0
```

The 0 in the lower right position is the remainder when the original polynomial is divided by $r - 2$. The quotient is $r^4 - 5r^2 - 12r + 28$, its coefficients found in the bottom row.

Since $r = 2$ appears to have multiplicity 3, we run through this process again, each time on the smaller quotient:

```
2 | 1   0  -5  -12   28
  |     2   4   -2  -28
  -------------------------
    1   2  -1  -14  | 0
```

One more time:

```
2 | 1   2  -1   14
  |     2   8   14
  -------------------
    1   4   7  | 0
```

The quotient is a quadratic, $r^2 + 4r + 7$. Thus, we have factored the original polynomial as $(r - 2)^3(r^2 + 4r + 7)$. This is a fifth-degree polynomial but only three of its roots (the 2, multiplicity 3) appears on the graph. Where are the other two? They are complex roots, found by the last factor. These will not appear on an xy-axis system. Using the quadratic formula on the last factor $r^2 + 4r + 7$, its roots are $r = -2 \pm i\sqrt{3}$.

The general solution is

$$y = C_1 e^{2x} + C_2 x e^{2x} + C_3 x^2 e^{2x} + C_4 e^{-2x} \cos\sqrt{3}x + C_5 e^{-2x} \sin\sqrt{3}x.$$

See an error? Have a suggestion?
Please visit www.surgent.net/debook

Section 14
Reduction of Order

Given a solution of a linear, homogeneous second-order ODE, it is possible to find another solution using a technique called **reduction of order**. The coefficients of y or its derivatives do not necessarily have to be constants.

Assume that the ODE has the form $a(x)y'' + b(x)y' + c(x)y = 0$ and that $y_1(x)$ is a solution.

We define the other solution $y_2(x) = v(x)y_1(x)$, where $v(x)$ is to be determined. Using the product rule, we find $y_2'(x)$ and $y_2''(x)$:

$$y_2'(x) = v(x)y_1'(x) + v'(x)y_1(x),$$

$$y_2''(x) = v(x)y_1''(x) + 2v'(x)y_1'(x) + v''(x)y_1(x).$$

These are substituted into the differential equation and simplified. For the sake of space, we write $y_2 = vy_1$, $y_2' = vy_1' + v'y_1$ and $y_2'' = vy_1'' + 2v'y_1' + vy_1''$. We also write a in place of $a(x)$, b in place of $b(x)$ and c in place of $c(x)$:

$$a(vy_1'' + 2v'y_1' + v''y_1) + b(vy_1' + v'y_1) + cvy_1 = 0$$

Distribute to clear parentheses and regroup according to degrees of v:

$$v''[ay_1] + v'[2ay_1' + by_1] + v\underbrace{[ay_1'' + by_1' + cy_1]}_{0} = 0$$

Notice that the expression multiplied to v is 0. This is because y_1 is a solution of the original differential equation so this expression is 0. We are left with

$$v''[ay_1] + v'[2ay_1' + by_1] = 0.$$

Replacing v' with u (and v'' with u'), we have a first-order ODE in terms of u. Thus, we have reduced the order to a single-order ODE which can be solved using separation of variables or integration factors.

$$u'[ay_1] + u[2ay_1' + by_1] = 0.$$

Remember, once u is determined, integrate one more time to get v.

Example 14.1: Let $y_1 = e^{2x}$ be one solution of $y'' - 4y' + 4y = 0$. Use reduction of order to find another solution.

Solution: Using the formula $u'[ay_1] + u[2ay_1' + by_1] = 0$, we have $a = 1$, $b = -4$, $y_1 = e^{2x}$ and $y_1' = 2e^{2x}$. Thus, we substitute:

$$u'[(1)(e^{2x})] + u[2(1)(2e^{2x}) + (-4)(e^{2x})] = 0.$$

In this example, the expression $2(1)(2e^{2x}) + (-4)(e^{2x}) = 0$, so we have

$$e^{2x}u' = 0.$$

Thus, $u' = 0$, so integrating, $u = k_1$, a constant. But since $v' = u$, we integrate again to find v, getting $v = k_1 x + k_2$. Since $y_2 = vy_1$, we have $y_2 = (k_1 x + k_2)e^{2x}$.

Since $y_1 = e^{2x}$ and $y_2 = (k_1 x + k_2)e^{2x}$ are solutions of $y'' - 4y' + 4y = 0$, the general solution is

$$y = C_1 y_1 + C_2 y_2$$
$$= C_1 e^{2x} + C_2 (k_1 x + k_2)e^{2x}.$$

Clearing parentheses, we have

$$y = C_1 e^{2x} + C_2 k_1 x e^{2x} + C_2 k_2 e^{2x}.$$

We can combine the first and third term, renaming $C_1 + C_2 k_2$ as "new" C_1, and $C_2 k_1$ as "new" C_2. Thus, the general solution can be written

$$y = C_1 e^{2x} + C_2 x e^{2x}.$$

You may recall that the auxiliary polynomial, $r^2 - 4r + 4 = 0$ has root $r = 2$, multiplicity 2. One solution is $y_1 = e^{2x}$ and the other $y_2 = xe^{2x}$. This process "justifies" that extra x.

To check linear independence, we find the Wronskian of the two components of the solution:

$$W(e^{2x}, xe^{2x}) = \det \begin{bmatrix} e^{2x} & xe^{2x} \\ 2e^{2x} & 2xe^{2x} + e^{2x} \end{bmatrix}$$
$$= e^{2x}(2xe^{2x} + e^x) - 2xe^{2x}(e^{2x})$$
$$= 2xe^{4x} + e^{4x} - 2xe^{4x} = e^{4x}.$$

The expression e^{4x} is never 0, for all x in the domain. Thus, the two functions are linearly independent.

The differential equations in Examples 14.2 and 14.3 are called Cauchy-Euler equations and are discussed in Section 17.

Example 14.2: Given that $y_1 = t^2$ is one solution of $t^2 y'' + \frac{t}{2} y' - 3y = 0$. Find another solution y_2 of this differential equation, show that y_1 and y_2 are linearly independent, and state the general solution.

Solution: Set $y_2 = vt^2$, where v represents function $v(t)$, and $u(t) = v'(t)$. Using the form $u'[ay_1] + u[2ay_1' + by_1] = 0$, we have $a = t^2$, $b = t/2$, $y_1 = t^2$ and $y_1' = 2t$. Substituting, we have

$$u'[(t^2)(t^2)] + u\left[2(t^2)(2t) + \left(\frac{t}{2}\right)(t^2)\right] = 0.$$

Simplifying, we have

$$t^4 u' + \frac{9}{2} t^3 u = 0.$$

Divide through by t^4 and note that $t \neq 0$:

$$u' + \frac{9}{2t} u = 0, \ t \neq 0.$$

This can be solved using an integration factor: $\mu(t) = e^{\int (9/2t)\, dt} = e^{(9/2)\ln t} = e^{\ln t^{9/2}} = t^{9/2}$. Thus,

$$u(t) = \frac{\int (t^{9/2})(0)\, dt + C}{t^{9/2}} = C t^{-9/2}.$$

But remember, we want $v(t)$, where $v'(t) = u(t)$. Thus, integrate once to find $v(t)$:

$$v(t) = \int C t^{-9/2}\, dt = -\frac{2}{7} C t^{-7/2} + D.$$

Since $v(t) = -\frac{2}{7} C t^{-7/2} + D$, then

$$y_2(t) = v(t) y_1(t) = \left(-\frac{2}{7} C t^{-7/2} + D\right) t^2 = -\frac{2}{7} C t^{-3/2} + D t^2.$$

The general solution is

$$y = C_1 y_1(t) + C_2 y_2(t)$$
$$= C_1 t^2 + C_2 \left(-\frac{2}{7} C t^{-3/2} + D t^2\right)$$
$$= C_1 t^2 - \frac{2}{7} C_2 C t^{-3/2} + C_2 D t^2.$$

Combine the first and third terms $C_1 t^2 + C_2 D t^2$ as $C_1 t^2$, where "new" $C_1 = C_1 + C_2 D$, and let "new" $C_2 = -\frac{2}{7} C_2 C$, so that the general solution is:

$$y = C_1 t^2 + C_2 t^{-3/2}.$$

Check: Let's check that $y_2 = t^{-3/2}$ solves $t^2 y'' + \frac{t}{2} y' - 3y = 0$.

Taking derivatives, we have $y_2' = -\frac{3}{2} t^{-5/2}$ and $y_2'' = \frac{15}{4} t^{-7/2}$. Substituting then simplifying, we have

$$t^2 \left(\frac{15}{4} t^{-7/2}\right) + \frac{t}{2} \left(-\frac{3}{2} t^{-5/2}\right) - 3(t^{-3/2}) = 0$$

$$\frac{15}{4} t^{-3/2} - \frac{3}{4} t^{-3/2} - 3 t^{-3/2} = 0$$

$$t^{-3/2} \underbrace{\left(\frac{15}{4} - \frac{3}{4} - 3\right)}_{0} = 0.$$

Now, we check that $y_1(t) = t^2$ and $y_2(t) = t^{-3/2}$ are linearly independent by finding its Wronskian:

$$W(t^2, t^{-3/2}) = \det \begin{bmatrix} t^2 & t^{-3/2} \\ 2t & (-3/2) t^{-5/2} \end{bmatrix}$$

$$= -\frac{3}{2} t^{-1/2} - 2 t^{-1/2}$$

$$= -\frac{7}{2} t^{-1/2}.$$

The expression $-\frac{7}{2} t^{-1/2}$ is never 0. This means that the two functions $y_1(t) = t^2$ and $y_2(t) = t^{-3/2}$ are linearly independent and that $y = C_1 t^2 + C_2 t^{-3/2}$, $t \neq 0$ is the general solution of $t^2 y'' + \frac{t}{2} y' - 3y = 0$.

Example 14.3: Given that $y_1 = x^3$ is a solution of $x^2 y'' - 5xy' + 9y = 0$, where $x > 0$, find another solution of this differential equation and state the general solution.

Solution: Using the form $u'[ay_1] + u[2ay_1' + b(t)y_1] = 0$, we have $a = x^2$, $b = -5x$, $y_1 = x^3$ and $y_1' = 3x^2$. Substituting, we have

$$u'[(x^2)(x^3)] + u[2(x^2)(3x^2) + (-5x)(x^3)] = 0$$
$$x^5 u' + (6x^4 - 5x^4)u = 0$$
$$x^5 u' + x^4 u = 0$$
$$u' + \frac{1}{x}u = 0, \quad x \neq 0.$$

Using the separation of variables method, we have

$$\frac{du}{u} = -\frac{dx}{x}$$

$$\int \frac{du}{u} = \int -\frac{dx}{x}$$

$$\ln u = -\ln x$$

$$u = \frac{1}{x}, \quad x > 0.$$

Since $u = v'$, integrate one more time to find v:

$$v = \int \frac{1}{x} dx = \ln x.$$

Thus, the second solution is

$$y = x^3 \ln x.$$

Check: We have $y' = x^2 + 3x^2 \ln(x)$ and $y'' = 5x + 6x \ln(x)$. Substituting:

$$x^2 \overbrace{(5x + 6x \ln x)}^{y''} - 5x \overbrace{(x^2 + 3x^2 \ln x)}^{y'} + 9 \overbrace{(x^3 \ln x)}^{y} = 0$$
$$5x^3 + 6x^3 \ln x - 5x^3 - 15x^3 \ln x + 9x^3 \ln x = 0$$
$$\underbrace{(6x^3 - 15x^3 + 9x^3)}_{0} \ln x + \underbrace{5x^3 - 5x^3}_{0} = 0, \quad \text{true.}$$

The general solution is

$$y = C_1 x^3 + C_2 x^3 \ln x, \quad x > 0.$$

This differential equation, when solved using the methods shown in Section 17, results in a repeated root. The second solution, linearly independent from the first, is what is seen above. See Example 17.4 for further discussion on this problem.

Example 14.4: Given that $y_1 = \cos 2x$ is one solution of $y'' + 4y = 0$, find another solution of this differential equation and state the general solution.

Solution: Once again, using the form $u'[ay_1] + u[2ay_1' + by_1] = 0$, we have $a = 1, b = 0, y_1 = \cos 2x$ and $y_1' = -2 \sin 2x$. Substituting, we have,

$$u'[(1)(\cos 2x)] + u[2(1)(-2\sin 2x) + (0)(\cos 2x)] = 0.$$

Simplified:

$$u' \cos 2x - 4u \sin 2x = 0.$$

Divide through by $\cos(2x)$:

$$u' - 4u \tan 2x = 0.$$

Separate the variables:

$$\frac{du}{u} = 4 \tan 2x \, dx.$$

Integrate:

$$\int \frac{du}{u} = 4 \int \frac{\sin 2x}{\cos 2x} dx$$

$$\ln u = -2 \ln(\cos 2x)$$

$$\ln u = \ln(\cos 2x)^{-2}$$

$$u = \cos^{-2} 2x = \sec^2 2x.$$

Thus,

$$v = \int u' = \int \sec^2 2x = \frac{1}{2} \tan 2x.$$

Therefore,

$$y_2 = vy_1 = \left(\frac{1}{2} \tan 2x\right) \cos 2x = \frac{1}{2}\left(\frac{\sin 2x}{\cos 2x}\right) \cos 2x = \frac{1}{2} \sin 2x \ .$$

When composed into the general solution, the leading $\frac{1}{2}$ is absorbed by the generic constant. We have

$$y = C_1 y_1 + C_2 y_2 = C_1 \cos 2x + C_2 \sin 2x.$$

Section 15
Undetermined Coefficients

Consider a linear, nth-order ODE with constant coefficients that is **not** homogeneous—that is, its forcing function is not 0. We can determine a general solution by using the **Method of Undetermined Coefficients**.

The usual routine is to find the general solution for the homogeneous case (call it y_h), then find a particular solution for the non-zero forcing function (call it y_p). The general solution is the sum: $y = y_h + y_p$.

Example 15.1: Find the general solution of $y'' - 3y' + 2y = 10e^{4x}$.

Solution: We first find the solution of the homogeneous case. The auxiliary polynomial is $r^2 - 3r + 2 = 0$, which factors as $(r-1)(r-2) = 0$. Thus, it has roots $r = 1$ and $r = 2$, and the homogeneous solution is

$$y_h = C_1 e^x + C_2 e^{2x}.$$

Now we find a particular solution for $y'' - 3y' + 2y = 10e^{4x}$.

We "guess" that it probably has the appearance $y_p = Ae^{4x}$. Taking derivatives, we have $y_p' = 4Ae^{4x}$ and $y_p'' = 16Ae^{4x}$. These are substituted:

$$(16Ae^{4x}) - 3(4Ae^{4x}) + 2(Ae^{4x}) = 10e^{4x}$$
$$16Ae^{4x} - 12Ae^{4x} + 2Ae^{4x} = 10e^{4x}$$
$$(16A - 12A + 2A)e^{4x} = 10e^{4x}$$
$$6Ae^{4x} = 10e^{4x}.$$

Therefore, $A = \frac{5}{3}$, and the particular solution is

$$y_p = \frac{5}{3} e^{4x}.$$

The general solution is the sum of the homogeneous and particular solutions:

$$y = y_h + y_p = C_1 e^x + C_2 e^{2x} + \frac{5}{3} e^{4x}.$$

The homogeneous solution is included because it has the effect of adding 0 to the particular solution when evaluated into the differential equation.

> **Important:** the particular solution must be in a form that is linearly independent of the terms of the homogeneous solution. In the above example, the form of the particular solution, e^{4x}, will be linearly independent of both terms, e^x and e^{2x}, of the homogeneous solution.

What happens if the form for the particular solution appears to be linearly dependent of the homogeneous solution's terms? The next example explores this case.

Example 15.2: Find the general solution of $y'' + 2y' - 15y = 6e^{3x}$.

Solution: The homogeneous solution is found first. The auxiliary polynomial is $r^2 + 2r - 15 = 0$, factoring to $(r+5)(r-3) = 0$, providing roots $r = -5$ and $r = 3$. Thus, $y_h = C_1 e^{-5x} + C_2 e^{3x}$ is the homogeneous solution.

Note that the forcing function, $6e^{3x}$, is *not* linearly independent of the homogeneous solution. It is a multiple of $C_2 e^{3x}$. Thus, we cannot forecast that the solution form has the form $y_p = Ae^{3x}$. Instead, we try $y_p = Axe^{3x}$.

Taking derivatives, we have $y_p' = 3Axe^{3x} + Ae^{3x}$ and $y_p'' = 9Axe^{3x} + 6Ae^{3x}$. These are substituted into the differential equation

$$(9Axe^{3x} + 6Ae^{3x}) + 2(3Axe^{3x} + Ae^{3x}) - 15(Axe^{3x}) = 6e^{3x}.$$

This simplifies to

$$9Axe^{3x} + 6Ae^{3x} + 6Axe^{3x} + 2Ae^{3x} - 15Axe^{3x} = 6e^{3x}.$$

In the next step, group the terms containing xe^{3x} or e^{3x}:

$$xe^{3x}(9A + 6A - 15A) + e^{3x}(6A + 2A) = 6e^{3x}.$$

Note that $9A + 6A - 15A = 0$, so that the first term vanishes. We have:

$$e^{3x}(6A + 2A) = 6e^{3x}, \quad \text{which simplifies to} \quad 8Ae^{3x} = 6e^{3x}$$

Thus, $8A = 6$ so that $A = \frac{3}{4}$. The particular solution to the forcing function is $y_p = \frac{3}{4}xe^{3x}$, and the general solution is

$$y = y_h + y_p = C_1 e^{-5x} + C_2 e^{3x} + \frac{3}{4}xe^{3x}.$$

Forms of the Particular Solution

The Method of Undetermined Coefficients requires one to make an educated guess to forecast the form of the particular solution. Below is a short guide to help in formulating the particular solution.

Assume the forcing function is linearly independent of the homogeneous solution.

If the forcing function is of the form e^{kx}, then choose $y_p = Ae^{kx}$. If e^{kx} is already present in the homogenous solution, then choose $y_p = Axe^{kx}$.

If the forcing function is of the form $\cos kx$ or $\sin kx$, then choose $y_p = A\cos kx + B\sin kx$. That is, choose *both* sine and cosine forms. If either $\sin kx$ or $\cos kx$ are already present in the homogenous solution, then choose $y_p = Ax\cos kx + Bx\sin kx$.

If the forcing function is an nth-degree polynomial, then choose the entire polynomial, $y_p = Ax^n + Bx^{n-1} + Cx^{n-2} + \cdots$.

Example 15.3: Find the general solution of $y'' - 4y' + 4y = 2\sin 3x$.

Solution: The homogeneous solution is $y_h = C_1 e^{2x} + C_2 x e^{2x}$.

The forcing function is linearly independent of e^{2x} and xe^{2x}. Thus, we choose

$$y_p = A\cos 3x + B\sin 3x.$$

The derivatives are

$$y_p' = -3A\sin 3x + 3B\cos 3x \quad \text{and} \quad y_p'' = -9A\cos 3x - 9B\sin 3x.$$

Substituted, we have

$$\underbrace{(-9A\cos 3x - 9B\sin 3x)}_{y_p''} - 4\underbrace{(-3A\sin 3x + 3B\cos 3x)}_{y_p'} + 4\underbrace{(A\cos 3x + B\sin 3x)}_{y_p} = 2\sin 3x.$$

Distribute to clear parentheses then group terms according to cos 3x and sin 3x:

$$(-9A - 12B + 4A)\cos 3x + (-9B + 12A + 4B)\sin 3x = 2\sin 3x$$

Simplified:

$$(-5A - 12B)\cos 3x + (12A - 5B)\sin 3x = 2\sin 3x.$$

There is no cosine term on the right side, which implies that $-5A - 12B = 0$. Comparing the sin 3x terms, we have $12A - 5B = 2$. This is a system:

$$-5A - 12B = 0$$
$$12A - 5B = 2.$$

Using any solution method, we get $A = \frac{24}{169}$ and $B = -\frac{10}{169}$. Thus, the particular solution is $y_p = \frac{24}{169}\cos 3x - \frac{10}{169}\sin 3x$, and the general solution is

$$y = C_1 e^{2x} + C_2 x e^{2x} + \frac{24}{169}\cos 3x - \frac{10}{169}\sin 3x.$$

Example 15.4: Find the general solution of $y'' + 2y' + 5y = t^2$.

Solution: The homogeneous solution is $y_h = C_1 e^{-t}\cos 2t + C_2 e^{-t}\sin 2t$. The forcing function t^2 is linearly independent of the terms in the homogenous solution. Thus, for the particular solution, we choose $y_p = At^2 + Bt + C$. Its derivatives are $y_p' = 2At + B$ and $y_p'' = 2A$, and the substitutions are made:

$$(2A) + 2(2At + B) + 5(At^2 + Bt + C) = t^2.$$

Distribute to clear parentheses:

$$2A + 4At + 2B + 5At^2 + 5Bt + 5C = t^2.$$

Rearrange by powers of t and equate to the term t^2, treating it as $1t^2 + 0t + 0$:

$$\underbrace{5At^2}_{1} + \underbrace{(4A + 5B)t}_{0} + \underbrace{(2A + 2B + 5C)}_{0} = 1t^2 + 0t + 0$$

The first equation gives $A = \frac{1}{5}$. From $4A + 5B = 0$, we obtain $B = -\frac{4}{5}A$, and since $A = \frac{1}{5}$, we have $B = -\frac{4}{5}\left(\frac{1}{5}\right) = -\frac{4}{25}$.

From $2A + 2B + 5C = 0$, we obtain $C = -\frac{2}{5}A - \frac{2}{5}B$. Making substitutions, we have $C = -\frac{2}{5}\left(\frac{1}{5}\right) - \frac{2}{5}\left(-\frac{4}{25}\right) = -\frac{2}{125}$.

The particular solution is $y_p = \frac{1}{5}t^2 - \frac{4}{25}t - \frac{2}{125}$, and the general solution is

$$y = C_1 e^{-t} \cos 2t + C_2 e^{-t} \sin 2t + \frac{1}{5}t^2 - \frac{4}{25}t - \frac{2}{125}.$$

Example 15.5: Find the general solution of $y'' + 4y = 5 \cos 2x$.

Solution: The homogeneous solution is $y_h = C_1 \cos 2x + C_2 \sin 2x$. The forcing function, $5 \cos 2x$, is not linearly independent of the terms in the homogeneous solution. Thus, we surmise the particular solution will be of the form $y_p = Ax \cos 2x + Bx \sin 2x$.

Find y'_p and y''_p:

$$y'_p = -2Ax \sin 2x + A \cos 2x + 2Bx \cos 2x + B \sin 2x.$$
$$y''_p = -4Ax \cos 2x - 4A \sin 2x - 4B \sin 2x + 4B \cos 2x.$$

The substitutions are made into the original differential equation:

$$\underbrace{-4Ax \cos 2x - 4A \sin 2x - 4B \sin 2x + 4B \cos 2x}_{y''_p} + 4\underbrace{(Ax \cos 2x + Bx \sin 2x)}_{y_p} = 5 \cos 2x.$$

Parentheses are cleared and terms grouped according to $x \cos 2x$, $x \sin(2x)$, $\cos 2x$ and $\sin 2x$, then coefficients are equated to the original forcing function, $5 \cos 2x$:

$$\overbrace{(-4A + 4A)}^{0} x \cos 2x + \overbrace{(-4B + 4B)}^{0} x \sin 2x + \overbrace{(4B)}^{5} \cos 2x + \overbrace{(-4A)}^{0} \sin 2x = 5 \cos 2x.$$

For the term $x \cos 2x$, we have $-4A + 4A = 0$ is true for all values of A, and for the term $x \sin 2x$, we have $-4B + 4B = 0$ for all values of B. This does not imply (yet) that $A = 0$ or $B = 0$. The values of A and B still remain unknown.

For the term $\cos 2x$, we have $4B = 5$, so that $B = \frac{5}{4}$, and for the term $\sin 2x$, we have $-4A = 0$, so that $A = 0$. Recall that our educated guess for the form of the particular solution was $y_p = Ax \cos 2x + Bx \sin 2x$. Now knowing A and B, the particular solution is $y_p = \frac{5}{4} x \sin 2x$.

The general solution is

$$y = C_1 \cos 2x + C_2 \sin 2x + \frac{5}{4} x \sin 2x.$$

Let's check the particular solution. We have $y_p' = \frac{5}{2} x \cos 2x + \frac{5}{4} \sin 2x$ and $y_p'' = -5x \sin 2x + \frac{5}{4} \cos 2x$. Substituting into the differential equation, we have

$$\underbrace{(-5x \sin 2x + 5 \cos 2x)}_{y_p''} + 4 \underbrace{\left(\frac{5}{4} x \sin 2x\right)}_{y_p} = 5 \cos 2x$$

$$-5x \sin 2x + 5 \cos 2x + 5x \sin 2x = 5 \cos 2x.$$

The first and third terms sum to 0, and the equation is true.

See an error? Have a suggestion?
Please visit www.surgent.net/debook

Section 16
Spring-Mass Systems

A common application of second-degree linear ODEs is a spring-mass system, in which an object with weight is attached to a spring, extending the spring and allowed to hang naturally coming to a rest state, then is compelled (for example, extended downward then let go) to bob up and down indefinitely. The system may be homogeneous if the mass bobs without any effects of the surrounding environment. If there is an external force imparting energy into the system (for example, it is shaken rhythmically), then the system will be non-homogeneous.

An object has weight, where weight is mass times gravity: $w = mg$. Here, g is the gravitational constant 32 ft/s². In these examples, a mass is usually pulled down then released to start the bobbing motion. Thus, the positive direction from rest state will be downward, which means the gravitational constant will be positive.

Hooke's Law: The force F_h to extend a spring a distance L feet is proportional to L, so that $F_h = kL$, where k is a constant of proportionality.

If the object is attached at the end of a spring, and is allowed to rest (not bob up and down), then the two forces cancel: the spring wants to contract in a direction opposite the weight, so that

$$mg - kL = 0, \quad \text{which gives} \quad mg = kL.$$

Suppose the object is pulled down and let go. It then starts to bob up and down. Let $u(t)$ be the displacement (in feet) at t seconds of the object from rest, where down is positive.

If the object originally extended the spring L feet, then the spring's length is also affected by the object's motion, so that $F_h = k(L + u(t))$.

The second derivative of displacement is $u''(t)$, which describes the object's acceleration at time t. Using the formula $F = ma$, we now have $F = mu''(t)$.

There may be a resistance force, F_d, defined as a force (in lbs) acting against the bobbing mass when the mass is at some speed, given by $u'(t)$. The force is proportional to the speed but in the opposite direction, so that $F_d = -\gamma u'(t)$.

The sum of all forces must equal $mu''(t)$:

$$mu''(t) = F_d + F_h$$
$$mu''(t) = mg - k(L + u(t)) - \gamma u'(t)$$
$$mu''(t) = mg - kL - ku(t) - \gamma u'(t)$$
$$mu''(t) = -ku(t) - \gamma u'(t) \quad \text{since } mg - kL = 0.$$

Collecting terms to one side, we have

$$mu''(t) + \gamma u'(t) + ku(t) = 0.$$

Remember, $w = mg$ so that $m = \frac{w}{g}$, and $k = \frac{\text{weight}}{\text{initial displacement}}$. Written out fully, the form of the equation is

$$\left(\frac{\text{weight}}{\text{gravity}}\right) u''(t) + \left(\frac{\text{resistance}}{\text{velocity}}\right) u'(t) + \left(\frac{\text{weight}}{\text{init. displacement}}\right) u(t) = 0.$$

The units are:

$$\left(\frac{\text{lbs}}{\text{ft/s}^2}\right)(\text{ft/s}^2) + \left(\frac{\text{lbs}}{\text{ft/s}}\right)(\text{ft/s}) + \left(\frac{\text{lbs}}{\text{ft}}\right)(\text{ft}).$$

Everything simplifies into pounds (lbs), which is a force.

Example 16.1: A mass weighing 4 lbs stretches a spring 2 inches (1/6 feet). The mass is pulled down 6 more inches (1/2 foot) then released. When the mass is moving at 3 feet/second, the surrounding medium applies a resistance force of 6 lbs. Find the initial value problem that governs the motion of the bobbing mass, and solve for $u(t)$.

Solution: Using the form

$$\left(\frac{\text{weight}}{\text{gravity}}\right) u''(t) + \left(\frac{\text{resistance}}{\text{velocity}}\right) u'(t) + \left(\frac{\text{weight}}{\text{init. displacement}}\right) u(t) = 0,$$

we have weight $w = 4$ lbs, gravity $g = 32$ ft/s², resistance = 6 lbs when velocity $v = 3$ ft/s, and initial displacement $1/6$ ft:

$$\left(\frac{4}{32}\right) u''(t) + \left(\frac{6}{3}\right) u'(t) + \left(\frac{4}{1/6}\right) u(t) = 0.$$

Simplified, we have

$$\frac{1}{8} u''(t) + 2u'(t) + 24u(t) = 0.$$

Multiplying by 8 to clear fractions, we have the initial-value ODE that models this system:

$$u''(t) + 16u'(t) + 192u(t) = 0, \quad \text{where} \quad u(0) = \frac{1}{2}, \ u'(0) = 0.$$

The initial conditions are $u(0) = \frac{1}{2}$, meaning at $t = 0$ seconds, the initial displacement was 6 inches (half a foot) and the original velocity is 0 ft/s, meaning it was simply released.

To solve for the general solution, the auxiliary polynomial is $r^2 + 16r + 192 = 0$. Using the quadratic formula, we have

$$r = \frac{-16 \pm \sqrt{16^2 - 4(1)(192)}}{2(1)} = \frac{-16 \pm \sqrt{-512}}{2} = \frac{-16 \pm 16i\sqrt{2}}{2} = -8 \pm 8i\sqrt{2}.$$

The general solution is

$$u(t) = e^{-8t}(C_1 \cos 8\sqrt{2}t + C_2 \sin 8\sqrt{2}t).$$

When $u(0) = 1/2$, the sine term vanishes, so $C_1 = 1/2$.

We now have

$$u(t) = e^{-8t}\left(\frac{1}{2}\cos 8\sqrt{2}t + C_2 \sin 8\sqrt{2}t\right).$$

To find C_2, we take the derivative of $u(t)$ and use the other initial condition, $u'(0) = 0$. The derivative is

$$u'(t) = e^{-8t}(-4\sqrt{2}\sin 8\sqrt{2}t + C_2 8\sqrt{2}\cos 8\sqrt{2}t)$$
$$- 8e^{-8t}\left(\frac{1}{2}\cos 8\sqrt{2}t + C_2 \sin 8\sqrt{2}t\right).$$

When $t = 0$, the sine terms vanish since $\sin 0 = 0$, the e^{8t} terms equal 1, and the cosine terms equal 1, since $\cos 0 = 1$. We have:

$$0 = C_2 8\sqrt{2} - 4, \quad \text{which gives} \quad C_2 = \frac{\sqrt{2}}{4}.$$

Thus, the equation that models the bobbing spring is

$$u(t) = e^{-8t}\left(\frac{1}{2}\cos 8\sqrt{2}t + \frac{\sqrt{2}}{4}\sin 8\sqrt{2}t\right).$$

Here is a graph of the bobbing spring over a 1-second interval of time:

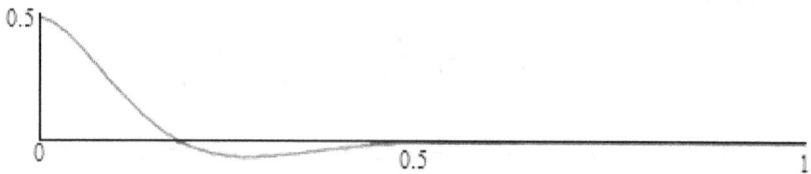

The spring bobs a couple times but quickly comes back to the rest state (the horizontal axis) due to the surrounding "dampening" force.

The mass is able to bob back to the rest state and beyond ("up and down"), but the amplitude trends to 0 as t increases. This is called a **damped** system. Note that the e^{-8t} factor in the solution acts as an "envelope", governing the amplitude. As t increases, e^{-8t} decreases to 0.

Example 16.2: A mass weighing 4 lbs stretches a spring 2 inches (1/6 feet). The mass is pulled down 6 more inches (1/2 foot) then released. There is no resistance force. Find the initial value problem that governs the motion of the bobbing mass and solve for $u(t)$. (This is the same example as the previous one, except without the resistance force)

Solution: The differential equation will lack the $u'(t)$ term. We have

$$\frac{1}{8}u''(t) + 24u(t) = 0, \quad \text{where} \quad u(0) = \frac{1}{2}, \ u'(0) = 0.$$

This simplifies to

$$u''(t) + 192u(t) = 0.$$

The auxiliary polynomial is $r^2 + 192 = 0$, which has roots $r = \pm 8i\sqrt{3}$. The general solution is

$$u(t) = C_1 \cos 8\sqrt{3}t + C_2 \sin 8\sqrt{3}t.$$

For the initial condition $u(0) = \frac{1}{2}$, we obtain $C_1 = 1$ since the sine term vanishes and the cosine term is 1. Differentiating, we have

$$u'(t) = -8\sqrt{3}\left(\frac{1}{2}\right) \sin 8\sqrt{3}t + 8\sqrt{3}C_2 \cos 8\sqrt{3}t.$$

For the initial condition $u'(0) = 0$. The sine term vanishes, the cosine at 0 gives 1, and we have $0 = 8\sqrt{3}C_2$, which implies that $C_2 = 0$. Thus, the solution is

$$u(t) = \frac{1}{2}\cos 8\sqrt{3}t.$$

Below is a 1-second graph of the bobbing mass. It never loses amplitude, bobbing forever.

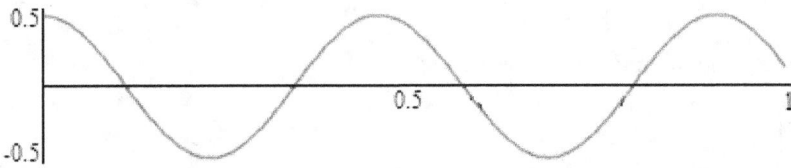

This is an **undamped** system.

Example 16.3: A mass weighing 4 lbs stretches a spring 2 inches (1/6 feet). The mass is pulled down 6 more inches (1/2 foot) then released. When the mass is moving at 3 feet/second, the surrounding medium applies a resistance force of 12 lbs. Find the initial value problem that governs the motion of the bobbing mass, and solve for $u(t)$. (This is the same example as the earlier example, but with the resistance force now double the original)

Solution: The differential equation is

$$\frac{1}{8}u''(t) + 4u'(t) + 24u(t) = 0.$$

Clearing fractions, we get

$$u''(t) + 32u'(t) + 192u(t) = 0, \quad \text{where} \quad u(0) = \frac{1}{2}, \; u'(0) = 0$$

The auxiliary polynomial is $r^2 + 32r + 192 = 0$, which is $(r + 24)(r + 8) = 0$, with solutions $r = -24$ and $r = -8$. The general solution is

$$u(t) = C_1 e^{-8t} + C_2 e^{-24t}.$$

For the initial condition $u(0) = \frac{1}{2}$, we have

$$\frac{1}{2} = C_1 + C_2.$$

Since $u'(t) = -8C_1 e^{-8t} - 24C_2 e^{-24t}$, incorporating the initial condition $u'(0) = 0$ results in

$$0 = -8C_1 - 24C_2.$$

Solving this system gives $C_1 = \frac{3}{4}$ and $C_2 = -\frac{1}{4}$. Thus, the solution is

$$u(t) = \frac{3}{4} e^{-8t} - \frac{1}{4} e^{-24t}.$$

One second of motion is shown below:

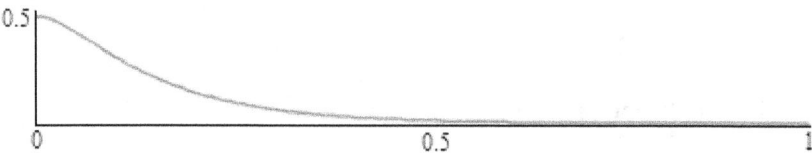

The mass never bobs. It just slowly moves back to rest state. This is an **overdamped** system.

There are three possible outcomes in a Spring-Mass System

The auxiliary polynomial has two complex conjugate solutions of the form $r = a \pm bi$. If $a = 0$, then the system is **undamped** and the mass bobs up and down forever. If $a < 0$, then the e^{at} factor of the solution acts as an "envelope" function, approaching 0 as t increases, and thus **damping** the bobbing nature of the mass. (Note that a is never positive in these problems since that would result in an envelope function that grows with time. The amplitude would increase, not decrease.)

The auxiliary polynomial has two real but different solutions. This is an **overdamped** system. The mass never bobs. It just slowly moves back to the rest state asymptotically.

The auxiliary polynomial has one real and repeated root. This is called a **critically damped** system.

To determine when a system is critically damped, solve the generic auxiliary polynomial form $mr^2 + \gamma r + k = 0$ using the quadratic formula, getting

$$r = \frac{-\gamma \pm \sqrt{\gamma^2 - 4mk}}{2m}.$$

To get a single root (multiplicity 2), set the discriminant to 0: $\gamma^2 - 4mk = 0$, which implies that $\gamma = 2\sqrt{mk}$. In the preceding examples, we have $m = 1/8$ and $k = 24$, so that $\gamma = 2\sqrt{(1/8)(24)} = 2\sqrt{3}$ would result in a critically damped system.

In the original example, the resistance was 6 lbs when velocity was 3 ft/s. If we set the resistance to $6\sqrt{3}$ lbs at a velocity of 3 ft/s, then we get $\gamma = \frac{6\sqrt{3}}{3} = 2\sqrt{3}$, which will cause the auxiliary polynomial to have just one root.

Example 16.4: A mass weighing 4 lbs stretches a spring 2 inches (1/6 feet). The mass is pulled down 6 more inches (1/2 foot) then released. When the mass is moving at 3 feet/second, the surrounding medium applies a resistance force of $6\sqrt{3}$ lbs. Find the initial value problem that governs the motion of the bobbing mass and solve for $u(t)$. (This is the same example as the earlier examples, but with the resistance force now chosen to enforce a critically-damped system)

Solution: Note that $\gamma = \frac{6\sqrt{3}}{3} = 2\sqrt{3}$. So that we get the differential equation $\frac{1}{8}u'' + 2\sqrt{3}u' + 24u = 0$ which has solution $r = -8\sqrt{3}$, multiplicity 2. The solution (with the same initial conditions as before) is

$$u(t) = e^{-8\sqrt{3}t}\left(4\sqrt{3}t + \frac{1}{2}\right).$$

One second of motion is shown below:

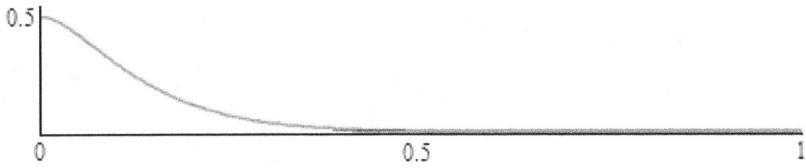

It does not look any different from an overdamped system. This is the "boundary" between the damped state and the overdamped state. If the resistance force is less than $6\sqrt{3}$ lbs, then we have a bobbing mass, and if the force is higher than $6\sqrt{3}$ lbs, then the mass does not bob.

An External Forcing Function

Suppose the original example has an external forcing function that imparts force into the system (for example, by shaking it rhythmically). Suppose the forcing function is $\sin(t/2)$. Thus, the differential equation is

Example 16.5: A mass weighing 4 lbs stretches a spring 2 inches (1/6 feet). The mass is pulled down 6 more inches (1/2 foot) then released. When the mass is moving at 3 ft/s, the surrounding medium applies a resistance force of 6 lbs. An external forcing function of $\sin(t/2)$ imparts energy into the system. Find the initial value problem that governs the motion of the bobbing mass and solve for $u(t)$.

Solution: The differential equation is now

$$u''(t) + 16u'(t) + 192u(t) = \sin(t/2), \quad u(0) = \frac{1}{2}, \quad u'(0) = 0.$$

The general solution is

$$u(t) = e^{-8t}\left(C_1 \cos(8\sqrt{2}t) + C_2 \sin(8\sqrt{2}t)\right) + 0.00521 \sin(t/2) - 0.000217 \cos(t/2).$$

After the initial conditions are incorporated, the specific solution is

$$u(t) = e^{-8t}\left(0.5 \cos(8\sqrt{2}t) + 0.354 \sin(8\sqrt{2}t)\right) + 0.00521 \sin(t/2) - 0.000217 \cos(t/2).$$

The coefficients have been rounded to three significant figures.

Here's the motion of the mass for the first 0.4 second:

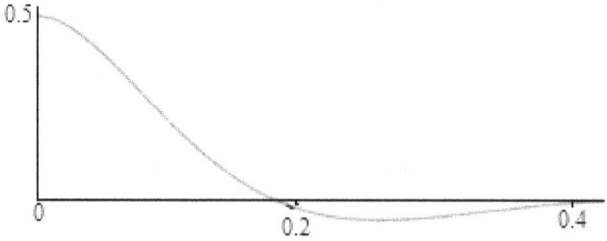

Then for the first 25 seconds. The forcing function "overwhelms" the natural decay of the unforced case, and the mass bobs according to the forcing function:

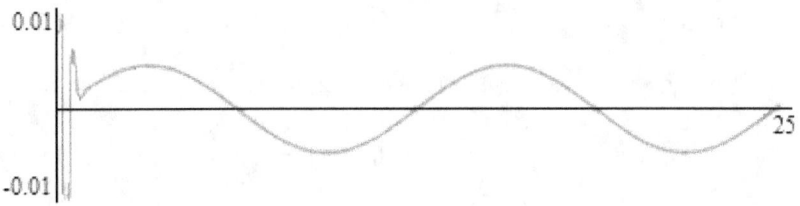

The amplitude is very low, about 0.005 units.

Section 17
Cauchy-Euler Equations

A Cauchy-Euler Equation has the form

$$0 = a_0 y + a_1 x y' + a_2 x^2 y'' + a_3 x^3 y^{(3)} + a_4 x^4 y^{(4)} + \cdots,$$

where a_n are constants. Typically, the equations are written in descending order, and the leading coefficient is divided out first. The remaining coefficients are simply renamed a_0, a_1 and so on. These equations are solved by assuming the solution has the form $y = x^n$.

Example 17.1: Solve $2x^2 y'' + 8xy' + 4y = 0$.

Solution: Divide by 2 to ensure a leading coefficient of 1:

$$x^2 y'' + 4xy' + 2y = 0.$$

Thus, $a_0 = 2$ and $a_1 = 4$.

Assuming $y = x^n$, we have $y' = nx^{n-1}$ and $y'' = n(n-1)x^{n-2}$. Make the substitutions:

$$\underbrace{n(n-1)x^{n-2}}_{y''} x^2 + \underbrace{4nx^{n-1}}_{y'} x + 2\underbrace{x^n}_{y} = 0.$$

Note that $x^{(n-2)} x^2 = x^n$ and $x^{(n-1)} x = x^n$. We now have

$$n(n-1)x^n + 4nx^n + 2x^n = 0.$$

Factor and simplify:

$$x^n[n(n-1) + 4n + 2] = 0$$
$$x^n[n^2 - n + 4n + 2] = 0$$
$$x^n[n^2 + 3n + 2] = 0.$$

We want $n^2 + 3n + 2 = 0$, and factoring gives $(n+2)(n+1) = 0$, so that $n = -2$ and $n = -1$. Thus, the two component solutions of the differential equation are x^{-2} and x^{-1}, and that the general solution is

$$y = C_1 x^{-2} + C_2 x^{-1}.$$

Because x is raised to negative powers, this requires $x \neq 0$.

Check: Taking derivatives, we have

$$y' = -2C_1 x^{-3} - C_2 x^{-2} \quad \text{and} \quad y'' = 6C_1 x^{-4} + 2C_2 x^{-3}.$$

Substituting into the original differential equations, we have

$$x^2 \underbrace{(6C_1 x^{-4} + 2C_2 x^{-3})}_{y''} + 4x \underbrace{(-2C_1 x^{-3} - C_2 x^{-2})}_{y'} + 2 \underbrace{(C_1 x^{-2} + C_2 x^{-1})}_{y} = 0.$$

Distribute to clear parentheses, then rearrange:

$$6C_1 x^{-2} + 2C_2 x^{-1} - 8C_1 x^{-2} - 4C_2 x^{-1} + 2C_1 x^{-2} + 2C_2 x^{-1} = 0$$
$$x^{-2} \underbrace{(6C_1 - 8C_1 + 2C_1)}_{0} + x^{-1} \underbrace{(2C_2 - 4C_2 + 2C_2)}_{0} = 0, \quad \text{true.}$$

To ensure that the two component solutions are linearly independent. We find the Wronskian. (Comment: if they were not linearly independent, we would add them together as like terms. But the Wronskian will confirm this)

$$W(x^{-2}, x^{-1}) = \det \begin{bmatrix} x^{-2} & x^{-1} \\ -2x^{-3} & -x^{-2} \end{bmatrix} = -x^{-4} + 2x^{-4} = x^{-4} = \frac{1}{x^4}.$$

This expression is never 0. Thus, the two component solutions are linearly independent.

Note: For second-order Cauchy-Euler equations, the substitution $y = x^n$ always results in the form

$$x^n[n^2 + (a_1 - 1)n + a_0] = 0.$$

Example 17.2: Solve $x^3 y^{(3)} - x^2 y'' - 10xy' + 10y = 0$.

Solution: Assuming $y = x^n$, we have $y' = nx^{n-1}$, $y'' = n(n-1)x^{n-2}$ and $y^{(3)} = n(n-1)(n-2)x^{n-3}$. Make the substitutions:

$$x^3(n(n-1)(n-2)x^{n-3}) - x^2(n(n-1)x^{n-2}) - 10x(nx^{n-1}) + 10(x^n) = 0.$$

As before, all expressions involving x become x^n and can be factored to the front and essentially set aside for now. We simplify the remaining expressions involving n and set its result to 0:

$$n(n-1)(n-2) - n(n-1) - 10n + 10 = 0$$
$$n^3 - 3n^2 + 2n - n^2 + n - 10n + 10 = 0$$
$$n^3 - 4n^2 - 7n + 10 = 0.$$

Using a graphing utility, the roots of this cubic polynomial are $n = 1$, $n = -2$ and $n = 5$. Thus, the general solution is

$$y = C_1 x + C_2 x^{-2} + C_3 x^5, \quad x \neq 0.$$

Example 17.3: Solve $x^2 y'' + 5xy' + 5y = 0$.

Solution: This is a second-order differential equation, so we use the short formula for these cases. We have $a_0 = 5$ and $a_1 = 5$. Simplified, the polynomial in terms of n is set to 0:

$$n^2 + 4n + 5 = 0.$$

The quadratic formula gives the roots:

$$n = \frac{-4 \pm \sqrt{4^2 - 4(5)}}{2} = \frac{-4 \pm 2i}{2} = -2 \pm i.$$

The roots are complex conjugates. The ostensible solution is

$$y = C_1 x^{-2+i} + C_2 x^{-2-i}.$$

For the time being, factor the x^{-2} to the front:

$$y = x^{-2}(C_1 x^i + x^{-i}).$$

We use the identity $e^{ib} = \cos b + i \sin b$ to simplify the x^i and x^{-i} terms. We do this by noting that $x = e^{\ln x}$ so that $x^i = e^{i \ln x}$ and $x^{-i} = e^{-i \ln x}$.

Thus,

$$x^i = \cos(\ln x) + i \sin(\ln x).$$

Doing the same for x^{-i} gives $\cos(\ln x) - i \sin(\ln x)$, but it is sufficient to perform this step for just one of the terms. See Example 12.1 for a detailed discussion.

The general solution is

$$y = C_1 x^{-2} \cos(\ln x) + C_2 x^{-2} \sin(\ln x).$$

In general, if the solution of the polynomial is $n = a \pm bi$, then the general solution is

$$y = C_1 x^a \cos(b \ln x) + C_2 x^a \sin(b \ln x), \quad x > 0.$$

In the above example $a = -2$ and $b = 1$.

Example 17.4: Solve $x^2 y'' - 5xy' + 9y = 0$.

Solution: We have $a_0 = -5$ and $a_1 = 9$. Simplified, the polynomial in terms of n is set to 0:

$$n^2 - 6n + 9 = 0$$
$$(n-3)(n-3) = 0$$
$$(n-3)^2 = 0.$$

Therefore, $n = 3$ is a root of multiplicity 2, and one solution is $y_1 = x^3$. Unfortunately, this method does not account for roots of higher multiplicity. The other solution is found by reduction of order and will be $y_2 = x^3 \ln x$. (See Example 14.3 for the details how the $\ln x$ was determined.)

Thus, the general solution is

$$y = C_1 x^3 + C_2 x^3 \ln x, \quad x > 0.$$

In Cauchy-Euler equations, the usual restriction is $x \neq 0$, but if the natural logarithm is present, then $x > 0$. However, it is possible to allow for negative x values by writing $\ln|x|$ in place of $\ln x$.

So that you don't embarrass yourself in front of your professor, Cauchy-Euler is pronounced ko-she-oiler.

Section 18
Laplace Transforms

Laplace transforms offer an alternative method to solve linear ordinary differential equations with initial conditions, with the following advantages over the previous methods we have discussed:

The process is formulaic and easily generalizable, it accounts for the initial conditions early in the process, not after the general solution has been determined, and it handles discontinuous cases and "pulse" cases more effectively.

Let $y = f(t)$ be a function. Its **Laplace Transform**, $H(s) = L\{f(t)\}(s)$, is a function in variable s, defined by

$$H(s) = L\{f(t)\}(s) = \int_0^\infty f(t)e^{-st}\, dt.$$

For convenience, we often write $H(s) = L\{f(t)\}$, dropping the (s), since it is understood to be present.

Case 1 (Constants): Let $f(t) = c$, where c is any constant. Then

$$H(s) = L\{c\} = \int_0^\infty ce^{-st}\, dt = c\int_0^\infty e^{-st}\, dt.$$

The integral $\int_0^\infty e^{-st}\, dt$ is found using limits:

$$\lim_{b\to\infty}\int_0^b e^{-st}\, dt = \lim_{b\to\infty}\left[-\frac{1}{s}e^{-st}\right]_0^b = \lim_{b\to\infty}\left[-\frac{1}{s}(e^{-sb} - (e^0))\right] = \frac{1}{s}.$$

Note that $\lim_{b\to\infty} e^{-sb} = 0$ in the second-to-last step above. Thus, when $f(t) = c$, its Laplace Transform is $H(s) = L\{c\} = \frac{c}{s}$.

Case 2 (exponential, base-e): Let $f(t) = e^{at}$. Then

$$H(s) = L\{e^{at}\} = \int_0^\infty e^{at}e^{-st}\, dt = \int_0^\infty e^{(a-s)t}\, dt.$$

The integral is evaluated using limits:

$$\lim_{b\to\infty}\int_0^b e^{(a-s)t}\, dt = \lim_{b\to\infty}\left[\frac{1}{a-s}e^{(a-s)t}\right]_0^b = \lim_{b\to\infty}\left[\frac{1}{a-s}\left(e^{(a-s)b} - (e^0)\right)\right].$$

When $s > a$, then $\lim_{b \to \infty} \left[e^{(a-s)b} \right] = 0$ since the coefficient of b is negative. Also, $e^0 = 1$. This simplifies to $\frac{1}{a-s}(0-1) = \frac{1}{s-a}$. Thus, when $f(t) = e^{at}$, its Laplace Transform is $H(s) = L\{e^{at}\} = \frac{1}{s-a}$.

Normally, it is not necessary to calculate the Laplace Transform each time. The table below lists the Laplace Transform for common functions. **These should be memorized.**

Function $y = f(t)$	Laplace Transform $H(s) = L\{f(t)\}$
$f(t) = c$	$H(s) = L\{c\} = \dfrac{c}{s}$
$f(t) = e^{at}$	$H(s) = L\{e^{at}\} = \dfrac{1}{s-a}$
$f(t) = t^n$	$H(s) = L\{t^n\} = \dfrac{n!}{s^{n+1}}$
$f(t) = \cos bt$	$H(s) = L\{\cos bt\} = \dfrac{s}{s^2 + b^2}$
$f(t) = \sin bt$	$H(s) = L\{\sin bt\} = \dfrac{b}{s^2 + b^2}$
$f(t) = e^{at} \cos bt$	$H(s) = L\{e^{at} \cos bt\} = \dfrac{s-a}{(s-a)^2 + b^2}$
$f(t) = e^{at} \sin bt$	$H(s) = L\{e^{at} \sin bt\} = \dfrac{b}{(s-a)^2 + b^2}$
$y' = f'(t)$	$H(s) = L\{y'\} = sL\{y\} - y(0)$
$y'' = f''(t)$	$H(s) = L\{y''\} = s^2 L\{y\} - sy(0) - y'(0)$

Reminder: $H(s) = L\{f(t)\}(s)$ is the more formal notation.

The last two rows show the Laplace Transform for the first and second derivatives of a function $y = f(t)$. The initial conditions $y(0) = f(0)$ and $y'' = f''(0)$ are incorporated into the process at this early stage.

Linearity of Laplace Transforms: The Laplace Transform operator can be distributed and any coefficients move to the front.

$$L\{c_1 f_1(t) + c_2 f_2(t)\} = c_1 L\{f_1(t)\} + c_2 L\{f_2(t)\}.$$

Example 18.1: Find $L\{t^4\}$.

Solution: Using the form $L\{t^n\} = \frac{n!}{s^{n+1}}$, we have

$$H(s) = L\{t^4\} = \frac{4!}{s^{4+1}} = \frac{24}{s^5}.$$

Example 18.2: Find $L\{e^{5t}\}$.

Solution: Using the form $L\{e^{at}\} = \frac{1}{s-a}$, we have

$$H(s) = L\{e^{5t}\} = \frac{1}{s-5}.$$

Example 18.3: Find $L\{\sin 7t\}$.

Solution: Using the form $L\{\sin bt\} = \frac{b}{s^2+b^2}$, we have

$$H(s) = L\{\sin 7t\} = \frac{7}{s^2+49}.$$

Example 18.4: Find $L\{e^{-2t} \cos 3t\}$.

Solution: Using the form $L\{e^{at} \cos bt\} = \frac{s-a}{(s-a)^2+b^2}$, we have

$$H(s) = L\{e^{-2t} \cos 3t\} = \frac{s+2}{(s+2)^2+9}.$$

Algebra may be necessary to calculate the Laplace Transform:

Example 18.5: Find $L\{(t+3)^2\}$.

Solution: Multiply the binomial: $L\{(t+3)^2\} = L\{t^2 + 6t + 9\}$. Now, using the linearity of the operator, we have

$$L\{(t+3)^2\} = L\{t^2 + 6t + 9\}$$
$$= L\{t^2\} + 6L\{t\} + L\{9\}$$
$$= \frac{2!}{s^{2+1}} + 6\left(\frac{1!}{s^{1+1}}\right) + 9\left(\frac{1}{s}\right)$$
$$= \frac{2}{s^3} + \frac{6}{s^2} + \frac{9}{s}.$$

Trigonometric identities may also be necessary:

Example 18.6: Find $L\{\sin^2 3t\}$.

Solution: We use the identity $\sin^2 t = \frac{1}{2} - \frac{1}{2}\cos 2t$. Thus,

$$L\{\sin^2 3t\} = L\left\{\frac{1}{2} - \frac{1}{2}\cos 6t\right\}$$
$$= L\left\{\frac{1}{2}\right\} - L\left\{\frac{1}{2}\cos 6t\right\}$$
$$= \frac{1}{2s} - \frac{1}{2}\left(\frac{s}{s^2 + 36}\right).$$

Inverting the Laplace Transform

As part of the solution process, we will need to invert the Laplace Transform, to find the function $f(t)$ such that

$$L^{-1}\{H(s)\} = f(t).$$

Some are easy to do by "inspection", as the following examples illustrate.

Example 18.7: Find $L^{-1}\left\{\frac{7}{s}\right\}$.

Solution: Since we know that $L\{c\} = \frac{c}{s}$, then

$$L^{-1}\left\{\frac{7}{s}\right\} = 7L^{-1}\left\{\frac{1}{s}\right\} = 7.$$

Example 18.8: Find $L^{-1}\left\{\frac{s}{s^2+25}\right\}$.

Solution: Since we know that $L\{\cos bt\} = \frac{s}{s^2+b^2}$, then $b = 5$ and we have

$$L^{-1}\left\{\frac{s}{s^2+25}\right\} = \cos 5t.$$

Example 18.9: Find $L^{-1}\left\{\frac{1}{s^6}\right\}$.

Solution: We know that $L\{t^n\} = \frac{n!}{s^{n+1}}$, so we conclude that $n = 5$ in this example. However, we need $5! = 120$ in the numerator to fully agree with the form. To do this, multiply inside by 120, and outside by $1/120$, and we have

$$L^{-1}\left\{\frac{1}{s^6}\right\} = \frac{1}{120}L^{-1}\left\{\frac{120}{s^6}\right\} = \frac{1}{120}t^5.$$

Example 18.10: Find $L^{-1}\left\{\frac{3}{s^2+21}\right\}$.

Solution: This is the form $L\{\sin bt\} = \frac{b}{s^2+b^2}$. The 3 in the numerator can be moved to the front, and from the denominator, we infer that since $b^2 = 21$, we must have $b = \sqrt{21}$.

However, the form needs the b value in the numerator. To do this, multiply inside by $\sqrt{21}$ and outside by $1/\sqrt{21}$:

$$L^{-1}\left\{\frac{3}{s^2+21}\right\} = 3L^{-1}\left\{\frac{1}{s^2+21}\right\} = 3\left(\frac{1}{\sqrt{21}}\right)L^{-1}\left\{\frac{\sqrt{21}}{s^2+21}\right\} = \frac{3}{\sqrt{21}}\sin\sqrt{21}t.$$

Rationalizing the denominator, an equivalent form is

$$L^{-1}\left\{\frac{3}{s^2+21}\right\} = \frac{\sqrt{21}}{7}\sin\sqrt{21}t.$$

Example 18.11: Find $L^{-1}\left\{\frac{s}{s^2+4s+9}\right\}$.

Solution: The denominator does not factor over the Reals, so we complete the square:

$$\frac{s}{s^2+4s+9} = \frac{s}{(s+2)^2+5}.$$

This form appears to be closest to $L\{e^{at}\cos bt\} = \frac{s-a}{(s-a)^2+b^2}$. We conclude that $a = -2$ and $b = \sqrt{5}$.

The numerator is not in the right form. We need the expression $s + 2$ in the numerator to agree with the form. So we add in 2 then subtract it back out, then split the numerator by grouping $s + 2$ and -2 as separate terms:

$$\frac{s}{(s+2)^2+5} = \frac{s+2-2}{(s+2)^2+5} = \frac{s+2}{(s+2)^2+5} - \frac{2}{(s+2)^2+5}.$$

The expression $\frac{s+2}{(s+2)^2+5}$ now matches the form for $L\{e^{at}\cos bt\} = \frac{s-a}{(s-a)^2+b^2}$. Thus,

$$L^{-1}\left\{\frac{s+2}{(s+2)^2+5}\right\} = e^{-2t}\cos\sqrt{5}t.$$

The expression $\frac{2}{(s+2)^2+5}$ almost matches the form for $L\{e^{at}\sin bt\} = \frac{b}{(s-a)^2+b^2}$. To match exactly, multiply inside by $\sqrt{5}$ and outside by $1/\sqrt{5}$, getting $\frac{2}{\sqrt{5}}\left(\frac{\sqrt{5}}{(s+2)^2+5}\right)$. Note that we moved the numerator 2 to the outside.

Performing the inversion, we have,

$$\frac{2}{\sqrt{5}}L^{-1}\left\{\frac{\sqrt{5}}{(s+2)^2+5}\right\} = \frac{2}{\sqrt{5}}e^{-2t}\sin\sqrt{5}t.$$

Combining everything, we obtain:

$$L^{-1}\left\{\frac{s}{s^2+4s+9}\right\} = L^{-1}\left\{\frac{s+2}{(s+2)^2+5}\right\} - \frac{2}{\sqrt{5}}L^{-1}\left\{\frac{\sqrt{5}}{(s+2)^2+5}\right\}$$

$$= e^{-2t}\cos\sqrt{5}t - \frac{2}{\sqrt{5}}e^{-2t}\sin\sqrt{5}t$$

$$= e^{-2t}\left(\cos\sqrt{5}t - \frac{2}{\sqrt{5}}\sin\sqrt{5}t\right).$$

Using the method of partial fractions is sometimes necessary:

Example 18.12: Find $L^{-1}\left\{\frac{1}{s(s^2+s-12)}\right\}$.

Solution: The denominator factors into linear factors:

$$\frac{1}{s(s^2+s-12)} = \frac{1}{s(s+4)(s-3)}.$$

We now use partial fraction decomposition:

$$\frac{1}{s(s+4)(s-3)} = \frac{A}{s} + \frac{B}{s+4} + \frac{C}{s-3}.$$

Recomposing, we have

$$\frac{1}{s(s+4)(s-3)} = \frac{A}{s} + \frac{B}{s+4} + \frac{C}{s-3}$$

$$= \frac{A(s+4)(s-3) + Bs(s-3) + Cs(s+4)}{s(s+4)(s-3)}.$$

Now, equate the two numerators:

$$1 = A(s+4)(s-3) + Bs(s-3) + Cs(s+4).$$

We find values for A, B and C by choosing "convenient" values for s:

If $s = 0$, then the terms containing B and C vanish. We have

$$1 = A(0 + 4)(0 - 3)$$
$$1 = -12A$$
$$A = -1/12.$$

If $s = -4$, then the A and C terms vanish, and we have

$$1 = B(-4)(-4 - 3)$$
$$1 = 28B$$
$$B = 1/28.$$

If $s = 3$, then the A and B terms vanish, and we have

$$1 = C(3)(3 + 4)$$
$$1 = 21C$$
$$C = 1/21.$$

Summarizing,

$$L^{-1}\left\{\frac{1}{s(s^2 + s - 12)}\right\} = L^{-1}\left\{\frac{-1/12}{s} + \frac{1/28}{s + 4} + \frac{1/21}{s - 3}\right\}.$$

Thus, performing the inversions, we have:

$$y = -\frac{1}{12} + \frac{1}{28}e^{-4t} + \frac{1}{21}e^{3t}.$$

Comment: *More discussion on partial fraction decomposition is found in an addendum at the end of this book.*

Section 19
Solving Initial Value Problems Using Laplace Transforms

We will now solve IVPs (Initial Value Problems) using the Laplace Transform. These usually involve a lot of algebra including decomposing rational expressions into partial fractions and solving linear systems.

Example 19.1: Solve $y'' + 2y' - 15y = 2t$, where $y(0) = 1$, $y'(0) = 0$.

Solution: Apply the Laplace Transform operator to both sides:

$$L\{y'' + 2y' - 15y\} = L\{2t\}.$$

By linearity, distribute the operator and move coefficients to the front:

$$L\{y''\} + 2L\{y'\} - 15L\{y\} = 2L\{t\}.$$

Expand the left side:

$$\underbrace{s^2 L\{y\} - sy(0) - y'(0)}_{L\{y''\}} + 2\underbrace{(sL\{y\} - y(0))}_{L\{y'\}} - 15L\{y\} = 2L\{t\}.$$

Note that this is the step in which the initial conditions are handled. Since $y(0) = 1$ and $y'(0) = 0$, we have

$$s^2 L\{y\} - s \cdot 1 - 0 + 2(sL\{y\} - 1) - 15L\{y\} = 2L\{t\}.$$

Simplify and distribute to clear parentheses:

$$s^2 L\{y\} - s + 2sL\{y\} - 2 - 15L\{y\} = 2L\{t\}.$$

On the right side, we have $L\{t\} = \frac{1}{s^2}$, so we have

$$s^2 L\{y\} - s + 2sL\{y\} - 2 - 15L\{y\} = \frac{2}{s^2}.$$

Now we isolate $L\{y\}$:

$$L\{y\}(s^2 + 2s - 15) = \frac{2}{s^2} + s + 2.$$

Get a common denominator on the right side:

$$L\{y\}(s^2 + 2s - 15) = \frac{s^3 + 2s^2 + 2}{s^2}.$$

Divide, and we have isolated $L\{y\}$:

$$L\{y\} = \frac{s^3 + 2s^2 + 2}{s^2(s^2 + 2s - 15)}.$$

The solution to the differential equation is found by inverting the right side. In other words,

$$y = L^{-1}\left\{\frac{s^3 + 2s^2 + 2}{s^2(s^2 + 2s - 15)}\right\}.$$

We need to "break apart" this rational expression into smaller summands. The denominator factors:

$$\frac{s^3 + 2s^2 + 2}{s^2(s^2 + 2s - 15)} = \frac{s^3 + 2s^2 + 2}{s^2(s + 5)(s - 3)}.$$

We need to decompose $\frac{s^3+2s^2+2}{s^2(s+5)(s-3)}$ using partial fractions. The factor s^2 is a linear factor of multiplicity 2 so it results in two fractional summands, while the factors $(s + 5)$ and $(s - 3)$ are each multiplicity 1, so they result in one summand each:

$$\frac{s^3 + 2s^2 + 2}{s^2(s + 5)(s - 3)} = \frac{A}{s} + \frac{B}{s^2} + \frac{C}{s + 5} + \frac{D}{s - 3}.$$

Please refer to the Partial Fraction Decomposition section at the end of this book for a discussion on how this decomposition is performed. We will get the following:

$$A = -\frac{4}{225}, B = -\frac{2}{15}, C = \frac{73}{200} \text{ and } D = \frac{47}{72}.$$

We have now separated the original rational expression into its four summands:

$$\frac{s^3 + 2s^2 + 2}{s^2(s + 5)(s - 3)} = \frac{-4/225}{s} + \frac{-2/15}{s^2} + \frac{73/200}{s + 5} + \frac{47/72}{s - 3}.$$

Thus, the solution of $y'' + 2y' - 15y = 2t$, $y(0) = 1$, $y'(0) = 0$ is

$$y = L^{-1}\left\{\frac{s^3 + 2s^2 + 2}{s^2(s+5)(s-3)}\right\}$$

$$= -\frac{4}{225}L^{-1}\left\{\frac{1}{s}\right\} - \frac{2}{15}L^{-1}\left\{\frac{1}{s^2}\right\} + \frac{73}{200}L^{-1}\left\{\frac{1}{s+5}\right\} + \frac{47}{72}L^{-1}\left\{\frac{1}{s-3}\right\}.$$

Recall that $L^{-1}\left\{\frac{1}{s}\right\} = 1$, $L^{-1}\left\{\frac{1}{s^2}\right\} = t$, $L^{-1}\left\{\frac{1}{s+5}\right\} = e^{-5t}$ and $L^{-1}\left\{\frac{1}{s-3}\right\} = e^{3t}$.
The coefficients we found simply move outside. The solution is:

$$y = -\frac{4}{225} - \frac{2}{15}t + \frac{73}{200}e^{-5t} + \frac{47}{72}e^{3t}.$$

Example 19.2: Solve the IVP $y'' + 2y' + 10y = t^2$, $y(0) = 1$, $y'(0) = -2$.

Solution: Apply the Laplace Transform operator to both sides and simplify:

$$L\{y''\} + 2L\{y'\} + 10L\{y\} = L\{t^2\}$$

$$L\{y''\} + 2L\{y'\} + 10L\{y\} = L\{t^2\}$$

$$s^2 L\{y\} - sy(0) - y'(0) + 2(sL\{y\} - y(0)) + 10L\{y\} = \frac{2!}{s^3}$$

$$s^2 L\{y\} - s + 2 + 2sL\{y\} - 2 + 10L\{y\} = \frac{2}{s^3}$$

$$L\{y\}(s^2 + 2s + 10) = \frac{s^4 + 2}{s^3}$$

$$L\{y\} = \frac{s^4 + 2}{s^3(s^2 + 2s + 10)}.$$

The solution is $y = L^{-1}\left\{\frac{s^4+2}{s^3(s^2+2s+10)}\right\}$.

Using partial fractions, the expression s^3 is a linear factor s with multiplicity 3, so it results in three partial fraction summands. The expression $s^2 + 2s + 10$ is an irreducible quadratic. Thus, the partial fraction decomposition is

$$\frac{s^4 + 2}{s^3(s^2 + 2s + 10)} = \frac{A}{s} + \frac{B}{s^2} + \frac{C}{s^3} + \frac{Ds + E}{s^2 + 2s + 10}.$$

Please refer to the Partial Fraction Decomposition section at the end of this book for a discussion on how this decomposition is performed. We will get the following:

$$A = -\frac{3}{250}, B = -\frac{1}{25}, C = \frac{1}{5}, D = \frac{253}{250} \text{ and } E = \frac{8}{125}.$$

Thus, the partial fraction decomposition is

$$\frac{s^4 + 2}{s^3(s^2 + 2s + 10)} = \frac{-\frac{3}{250}}{s} + \frac{-\frac{1}{25}}{s^2} + \frac{\frac{1}{5}}{s^3} + \frac{\frac{253}{250}s + \frac{8}{125}}{s^2 + 2s + 10}.$$

We'll concentrate on the last term for now. Complete the square:

$$\frac{\frac{253}{250}s + \frac{8}{125}}{s^2 + 2s + 10} = \frac{\frac{253}{250}s + \frac{8}{125}}{(s+1)^2 + 9}.$$

We intend to use the Laplace Transforms $L\{e^{at} \cos bt\} = \frac{s-a}{(s-a)^2+b^2}$ and $L\{e^{at} \sin bt\} = \frac{b}{(s-a)^2+b^2}$. Thus, we "need" an $(s+1)$ in the numerator. We do this by writing $s = s + 1 - 1$.

$$\frac{\frac{253}{250}s + \frac{8}{125}}{(s+1)^2 + 9} = \frac{\frac{253}{250}(s+1-1) + \frac{8}{125}}{(s+1)^2 + 9} = \frac{\frac{253}{250}(s+1) - \frac{253}{250} + \frac{8}{125}}{(s+1)^2 + 9}$$

$$= \frac{\frac{253}{250}(s+1) - \frac{237}{250}}{(s+1)^2 + 9} = \frac{\frac{253}{250}(s+1)}{(s+1)^2 + 9} + \frac{-\frac{237}{250}}{(s+1)^2 + 9}.$$

Assembling all the terms together, we have

$$\frac{s^4 + 2}{s^3(s^2 + 2s + 10)} = \frac{-\frac{3}{250}}{s} + \frac{-\frac{1}{25}}{s^2} + \frac{\frac{1}{5}}{s^3} + \frac{\frac{253}{250}(s+1)}{(s+1)^2 + 9} + \frac{-\frac{237}{250}}{(s+1)^2 + 9}.$$

Thus, $y = L^{-1}\left\{\frac{-\frac{3}{250}}{s} + \frac{-\frac{1}{25}}{s^2} + \frac{\frac{1}{5}}{s^3} + \frac{\frac{253}{250}(s+1)}{(s+1)^2+9} + \frac{-\frac{237}{250}}{(s+1)^2+9}\right\}.$

Distribute the inverse-Laplace operator and move constants to the front as coefficients. Balance with constants when necessary. The inversions are:

$$L^{-1}\left\{\frac{-\frac{3}{250}}{s}\right\} = -\frac{3}{250}L^{-1}\left\{\frac{1}{s}\right\} = -\frac{3}{250},$$

$$L^{-1}\left\{\frac{-\frac{1}{25}}{s^2}\right\} = -\frac{1}{25}L^{-1}\left\{\frac{1}{s^2}\right\} = -\frac{1}{25}t,$$

108

$$L^{-1}\left\{\frac{\frac{1}{5}}{s^3}\right\} = \frac{1}{5}L^{-1}\left\{\frac{1}{s^3}\right\} = \frac{1}{5}\cdot\frac{1}{2}L^{-1}\left\{\frac{2}{s^3}\right\} = \frac{1}{10}t^2,$$

$$L^{-1}\left\{\frac{\frac{253}{250}(s+1)}{(s+1)^2+9}\right\} = \frac{253}{250}L^{-1}\left\{\frac{s+1}{(s+1)^2+9}\right\} = \frac{253}{250}e^{-t}\cos 3t,$$

$$L^{-1}\left\{\frac{-\frac{237}{250}}{(s+1)^2+9}\right\} = -\frac{237}{250}\cdot\frac{1}{3}L^{-1}\left\{\frac{3}{(s+1)^2+9}\right\} = -\frac{79}{250}e^{-t}\sin 3t.$$

In the third term, we need a 2 in the numerator, so one is inserted (multiplied in) and balanced with a 1/2 multiplier. In the fifth term, we need a 3 in the numerator, and it is balanced with a 1/3 multiplier. Inverting each transform, we have the final result:

$$y = -\frac{3}{250} - \frac{1}{25}t + \frac{1}{10}t^2 + \frac{253}{250}e^{-t}\cos 3t - \frac{79}{250}e^{-t}\sin 3t.$$

Section 20
Laplace Transforms of Discontinuous Forcing Functions

We need a better way to describe functions with discontinuities. We introduce the **Heaviside Function**, which is

$$u_c(t) = \begin{cases} 0, & t < c \\ 1, & t \geq c \end{cases}$$

The graph of $u_1(t) = \begin{cases} 0, & t < 1 \\ 1, & t \geq 1 \end{cases}$ is below.

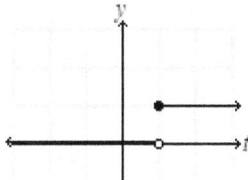

The Heaviside Function is "off" (= 0) when $t < c$, and "on" (= 1) when $t \geq c$. The jump occurs at $t = c$ and is always closed at the left endpoint of each jump, and open at any right endpoint.

Example 20.1: Sketch $y = 3u_2(t)$.

Solution: The function is rewritten in piecewise form:

$$y = 3u_2(t) = \begin{cases} 0, & t < 2 \\ 3, & t \geq 2 \end{cases}$$

The jump discontinuity occurs at $c = 2$ and has a vertical change of 3 units. Its graph is:

Note how much cleaner $3u_2(t)$ is in expressing the piecewise model.

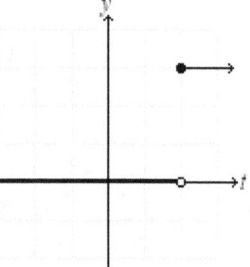

Example 20.2: Sketch $y = 1 - u_4(t)$.

Solution: The jump occurs at $c = 4$ and will have a vertical change of –1 unit.

When $t < 4$, then $u_4(t) = 0$, and $y = 1 - 0 = 1$.

When $t \geq 4$, then $u_4(t) = 1$, and $y = 1 - 1 = 0$.

The graph is:

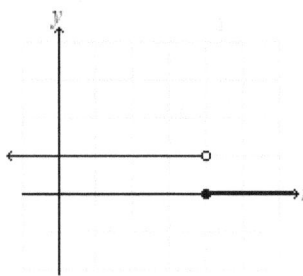

This is the same as the piece-wise defined function

$$y = \begin{cases} 1, & t < 4 \\ 0, & t \geq 4 \end{cases}$$

Example 20.3: Sketch $y = 2 - 3u_1(t) + 5u_3(t) - u_5(t)$.

Solution: There are three jumps, at $c = 1$, 3 and 5.

For $t < 1$, all three of the u terms are 0. Thus,

$$y = 2 - 3(0) + 5(0) - (0) = 2.$$

For $1 \leq t < 3$, we have $u_1(t) = 1$ but the other u terms are 0. Thus,

$$y = 2 - 3(1) + 5(0) - (0) = -1.$$

For $3 \leq t < 5$, we have $u_3(t) = 1$. Note that $u_1(t) = 1$ (it stays "on") but that $u_5(t) = 0$. Thus,

$$y = 2 - 3(1) + 5(1) - (0) = 2 - 3 + 5 = 4.$$

For $t \geq 5$, we have $u_5(t) = 1$. Note that $u_1(t) = 1$ and that $u_3(t) = 1$. All of the u terms are now "on". Thus,

$$y = 2 - 3(1) + 5(1) - (1) = 2 - 3 + 5 - 1 = 3.$$

Its graph is

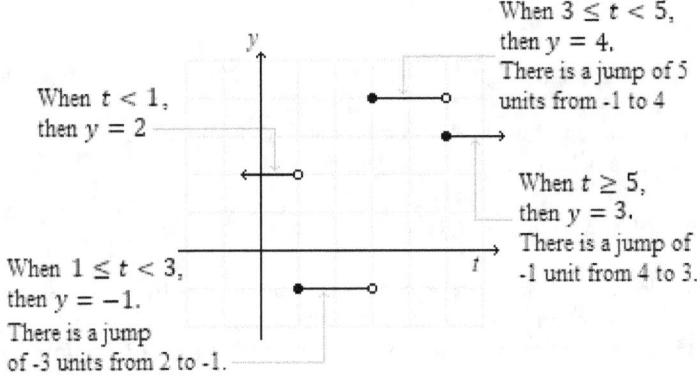

The Heaviside Function can be combined with common functions, acting as a coefficient, as the next example illustrates.

Example 20.4: Sketch $y = u_2(t)t^2$.

Solution: When $t < 2$, we have $u_2(t) = 0$, so that $y = 0t^2 = 0$. When $t \geq 2$, we have $u_2(t) = 1$, so that $y = 1t^2 = t^2$.

This is equivalent to the piecewise notation $y = \begin{cases} 0, & t < 2 \\ t^2, & t \geq 2 \end{cases}$.

The graph is 0 when $t < 2$, and then for $t \geq 2$, the parabola t^2 "starts" at (2,4) and continues upward.

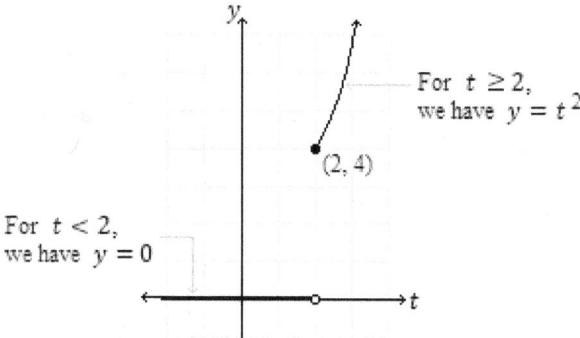

The $u_c(t)$ notation is useful for combining two or more function types into a single function, where continuity is desired.

Example 20.5: Sketch $y = u_2(t)(t-2)^2$.

Solution. When $t < 2$, we have $u_2(t) = 0$, so that $y = 0(t-2)^2 = 0$. When $t \geq 2$, we have $u_2(t) = 1$, so that $y = 1(t-2)^2 = (t-2)^2$. Here, the graph $(t-2)^2$ is a shift of t^2 two units to the right. It "starts" at (2,0). Note that is it continuous with the portion $y = 0$, where $t < 2$.

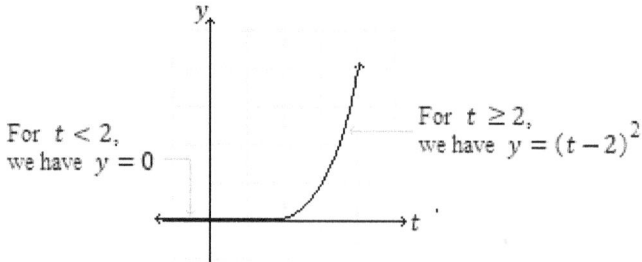

Laplace Transform of $y = u_c(t)f(t-c)$

Start with:

$$L\{u_c(t)f(t-c)\} = \int_0^\infty u_c(t)f(t-c)e^{-st}\,dt.$$

Since $u_c(t) = 0$ when $0 \le t < c$, and $u_c(t) = 1$ when $t \ge c$, we have

$$L\{u_c(t)f(t-c)\} = \int_c^\infty f(t-c)e^{-st}\,dt.$$
(The integral from 0 to c is 0)

Now, let $w = t - c$, so that $t = w + c$. This is a shift of c units to the right. Furthermore, $dw = d(t-c) = dt - dc = dt$, since $dc = 0$. We now have

$$\int_c^\infty f(t-c)e^{-st}\,dt = \int_0^\infty f(w)e^{-s(w+c)}\,dw.$$

Note that $e^{-s(w+c)} = e^{-sw-sc} = e^{-sw}e^{-sc}$, and that e^{-sc} is constant with respect to w. Thus, we have

$$\int_0^\infty f(w)e^{-s(w+c)}\,dw = e^{-sc}\int_0^\infty f(w)e^{-sw}\,dw.$$

The integral on the right is just the Laplace Transform of $f(w)$:

$$L\{f(w)\} = \int_0^\infty f(w)e^{-sw}\,dw.$$

Here, w is a dummy variable. It can be replaced with t. Thus, we have

$$L\{u_c(t)f(t-c)\} = e^{-sc}L\{f(t)\}.$$

For this to work, shift the function f by c units first, then evaluate the Laplace Transform as though the c was not present.

Example 20.6: Find $L\{u_3(t)\}$.

Solution: We can treat $u_3(t)$ as $u_3(t) = u_3(t) \cdot 1$. Shifting $y = 1$ left or right makes no difference as it is a horizontal line, so we can proceed:

$$L\{u_3(t)\} = e^{-3s}L\{1\} = e^{-3s}\left(\frac{1}{s}\right) = \frac{e^{-3s}}{s}.$$

Example 20.7: Find $L\{u_5(t)(t-5)^3\}$.

Solution: Here, $c = 5$, so the function t^3 must be shifted 5 units to the right, and we see that $(t-5)^3$ already has the shift accounted for. Thus, we have

$$L\{u_5(t)(t-5)^3\} = e^{-5s}L\{t^3\} = e^{-5s}\left(\frac{3!}{s^{3+1}}\right) = \frac{6e^{-5s}}{s^4}.$$

In the next example, the shift needs to be built in before the Laplace Transform can be performed:

Example 20.8: Find $L\{u_3(t)t^2\}$.

Solution: Here, we have $c = 3$, so we need to rewrite t^2 so that is has a shift of 3 units to the right. We do this by writing $t^2 = (t - 3 + 3)^2$.

Now, multiply by grouping $t - 3$ and 3 separately:

$$t^2 = (t - 3 + 3)^2 = (t-3)^2 + 2(t-3)(3) + 3^2$$

With a shift of 3 units to the right built in, $t^2 = (t-3)^2 + 6(t-3) + 9$. Thus,

$$L\{u_3(t)t^2\} = e^{-3s}L\{t^2 + 6t + 9\} = e^{-3s}\left(\frac{2}{s^3} + \frac{6}{s^2} + \frac{9}{s}\right).$$

Remember, once the shift is incorporated, take the Laplace Transform of each term as though the shift was not there. The leading e^{-3s} is a reminder that a shift has taken place.

Example 20.9: Find $L\{u_{\pi/4}(t)\sin t\}$.

Solution: Since $c = \frac{\pi}{4}$, we build in the shift by writing $\sin t$ as $\sin\left(t - \frac{\pi}{4} + \frac{\pi}{4}\right)$. Now, use the identity $\sin(a + b) = \sin a \cos b + \cos a \sin b$, where $a = t - \frac{\pi}{4}$ and $b = \frac{\pi}{4}$:

$$\sin\left(t - \frac{\pi}{4} + \frac{\pi}{4}\right) = \sin\left(t - \frac{\pi}{4}\right)\cos\left(\frac{\pi}{4}\right) + \cos\left(t - \frac{\pi}{4}\right)\sin\left(\frac{\pi}{4}\right).$$

Recall that $\cos\left(\frac{\pi}{4}\right) = \frac{\sqrt{2}}{2}$ and that $\sin\left(\frac{\pi}{4}\right) = \frac{\sqrt{2}}{2}$. We have

$$\sin\left(t - \frac{\pi}{4} + \frac{\pi}{4}\right) = \frac{\sqrt{2}}{2}\sin\left(t - \frac{\pi}{4}\right) + \frac{\sqrt{2}}{2}\cos\left(t - \frac{\pi}{4}\right).$$

Now the shifts are incorporated. We determine the Laplace Transform:

$$L\{u_{\pi/4}(t)\sin t\} = e^{-(\pi/4)s}L\left\{\frac{\sqrt{2}}{2}\sin\left(t - \frac{\pi}{4}\right) + \frac{\sqrt{2}}{2}\cos\left(t - \frac{\pi}{4}\right)\right\}$$

$$= \frac{\sqrt{2}}{2}e^{-(\pi/4)s}L\{\sin t + \cos t\}$$

$$= \frac{\sqrt{2}}{2}e^{-(\pi/4)s}\left[\left(\frac{1}{s^2 + 1}\right) + \left(\frac{s}{s^2 + 1}\right)\right]$$

$$= \frac{\sqrt{2}}{2}e^{-(\pi/4)s}\left(\frac{s + 1}{s^2 + 1}\right).$$

Inverting Laplace Transforms with $u_c(t)$ notation.

When there is a e^{-cs} factor in a Laplace Transform, this means there is a $u_c(t)$ in the resulting inversion (solution function), indicating the function is defined piecewise.

Example 20.10: Find $L^{-1}\left\{\frac{5e^{-6s}}{s^2}\right\}$.

Solution: The 5 can be moved out front. It is just a coefficient that has no bearing on the calculations. It simply rides along at each step:

$$L^{-1}\left\{\frac{5e^{-6s}}{s^2}\right\} = 5L^{-1}\left\{\frac{e^{-6s}}{s^2}\right\}.$$

The e^{-6s} suggests that there is a $u_6(t)$ in the solution function. Thus, we need to invert $\frac{1}{s^2}$. Recall that $L\{t\} = \frac{1}{s^2}$, so that $L^{-1}\left\{\frac{1}{s^2}\right\} = t$. In the inversion step, the shift needs to be present. Thus, we have

$$y = L^{-1}\left\{\frac{5e^{-6s}}{s^2}\right\} = 5u_6(t)(t-6).$$

Example 20.11: Find $L^{-1}\left\{\frac{e^{-2s}}{s^4}\right\}$.

Solution: The e^{-2s} suggests that there is a $u_2(t)$ in the solution function. The shift is $c = 2$ units. Ignoring the e factor for the moment, we find $L^{-1}\left\{\frac{1}{s^4}\right\}$. Recalling the general formula $L\{t^n\} = \frac{n!}{s^{n+1}}$, we surmise that $n = 3$, so we need $3! = 6$ in the numerator and its reciprocal outside:

$$L^{-1}\left\{\frac{1}{s^4}\right\} = \frac{1}{6}L^{-1}\left\{\frac{6}{s^4}\right\} = \frac{1}{6}t^3.$$

To perform the inversion, add in the shift of 2 units in the variable t. Thus,

$$y = L^{-1}\left\{\frac{e^{-2s}}{s^4}\right\} = \frac{1}{6}u_2(t)(t-2)^3.$$

Example 20.12: Find $L^{-1}\left\{\frac{e^{-s}}{s^2+9}\right\}$.

Solution: The e^{-s} suggests that there is a $u_1(t)$ is in the solution function and that $c = 1$ is the shift. Recall that

$$L\{\sin 3t\} = \frac{3}{s^2 + 9}.$$

Multiplying inside by 3 and outside by 1/3, we have,

$$L^{-1}\left\{\frac{1}{s^2+9}\right\} = \frac{1}{3}L^{-1}\left\{\frac{3}{s^2+9}\right\} = \frac{1}{3}\sin 3t.$$

The shift $c = 1$ is then accounted for in the sine function. Thus,

$$y = L^{-1}\left\{\frac{e^{-s}}{s^2+9}\right\} = \frac{1}{3}u_1(t)\sin(3(t-1)).$$

Section 21
Using Laplace Transforms to Solve IVPs with Piecewise Forcing Functions

We can now use Laplace Transforms to efficiently handle IVPs that have piecewise (continuous or discontinuous) forcing functions.

Example 21.1: Find the solution of the IVP

$$y'' + 2y' + 5y = \begin{cases} 0, & t < 4 \\ 1, & t \geq 4 \end{cases}, \quad y(0) = 1, \; y'(0) = -1.$$

Solution: Rewrite the forcing function using the $u_c(t)$ notation:

$$y'' + 2y' + 5y = u_4(t), \quad y(0) = 1, \; y'(0) = -1.$$

Now apply the Laplace Transform Operator to both sides and simplify:

$$L\{y''\} + 2L\{y'\} + 5L\{y\} = L\{u_4(t)\}$$

$$s^2 L\{y\} - s\underbrace{y(0)}_{1} - \underbrace{y'(0)}_{-1} + 2\left(sL\{y\} - \underbrace{y(0)}_{1}\right) + 5L\{y\} = L\{u_4(t)\}$$

$$s^2 L\{y\} - s + 1 + 2sL\{y\} - 2 + 5L\{y\} = \frac{e^{-4s}}{s}$$

$$L\{y\}[s^2 + 2s + 5] = \frac{e^{-4s}}{s} + s + 1.$$

Now isolate $L\{y\}$. A strategy in these cases is to have the e^{-4s} in one term, and all else in another term, and treat the problem as two smaller problems, to be combined at the end.

$$L\{y\} = \frac{e^{-4s}}{s(s^2 + 2s + 5)} + \frac{s+1}{s^2 + 2s + 5}.$$

The solution is the inversion of the above expressions:

$$y = L^{-1}\left\{\frac{e^{-4s}}{s(s^2 + 2s + 5)} + \frac{s+1}{s^2 + 2s + 5}\right\}.$$

For the term without the e^{-4s}, note that the denominator $s^2 + 2s + 5$ is an irreducible quadratic over the reals, so we complete the square:

$$\frac{s+1}{s^2+2s+5} = \frac{s+1}{(s+1)^2+4}.$$

This is exactly the form

$$L\{e^{at} \cos bt\} = \frac{s-a}{(s-a)^2+b^2},$$

Where $a = -1$ and $b = 2$. Thus,

$$L^{-1}\left\{\frac{s+1}{(s+1)^2+4}\right\} = e^{-t} \cos 2t.$$

This is one part of the general solution.

Now we find $L^{-1}\left\{\frac{e^{-4s}}{s(s^2+2s+5)}\right\}$. The e^{-4s} will result in $u_4(t)$ appearing in the solution function. We mentally note this fact, then essentially ignore it for the next few steps, as we decompose $\frac{1}{s(s^2+2s+5)}$ into smaller fractions using partial fraction decomposition:

$$\frac{1}{s(s^2+2s+5)} = \frac{A}{s} + \frac{Bs+C}{s^2+2s+5} = \frac{A(s^2+2s+5)+(Bs+C)s}{s(s^2+2s+5)}.$$

Equating the numerators, we have $1 = A(s^2 + 2s + 5) + (Bs + C)s$.

Collecting terms according to powers of s, we have

$$1 = \underbrace{(A+B)}_{0} s^2 + \underbrace{(2A+C)}_{0} s + \underbrace{5A}_{1}.$$

This forces $A + B = 0$, $2A + C = 0$ and $5A = 1$. Working backwards, we have $A = \frac{1}{5}$, then $B = -\frac{1}{5}$ and lastly, $C = -\frac{2}{5}$. So now we have

$$\frac{1}{s(s^2+2s+5)} = \frac{\frac{1}{5}}{s} - \frac{\frac{1}{5}s + \frac{2}{5}}{s^2+2s+5}.$$

Completing the square on the second term gives $s^2 + 2s + 5 = (s+1)^2 + 4$. Thus, we need to have $s+1$ in the numerator. Note that $s = s + 1 - 1$:

$$\frac{\frac{1}{5}s + \frac{2}{5}}{s^2 + 2s + 5} = \frac{\frac{1}{5}(s + 1 - 1) + \frac{2}{5}}{(s+1)^2 + 4} = \frac{\frac{1}{5}(s+1) + \frac{1}{5}}{(s+1)^2 + 4}$$

$$= \frac{\frac{1}{5}(s+1)}{(s+1)^2 + 4} + \underbrace{\frac{1}{2} \cdot \frac{\frac{1}{5}(2)}{(s+1)^2 + 4}}_{\substack{\text{A 2 and a 1/2 are} \\ \text{included here to} \\ \text{prepare this for} \\ \text{the sine form.}}}.$$

These forms are ready to be inverted. In both cases, note that $a = -1$ and $b = 2$, as before. The e^{-4s} is brought back into the equation:

$$L^{-1}\left\{\frac{e^{-4s}}{s(s^2 + 2s + 5)}\right\} = L^{-1}\left\{e^{-4s}\left(\frac{\frac{1}{5}}{s} - \frac{\frac{1}{5}(s+1)}{(s+1)^2 + 4} - \frac{1}{2} \cdot \frac{\frac{1}{5}(2)}{(s+1)^2 + 4}\right)\right\}.$$

For $L^{-1}\left\{\frac{1}{5} \cdot \frac{e^{-4s}}{s}\right\}$, we have

$$\frac{1}{5}u_4(t) \cdot 1 \quad \text{since } L^{-1}\left\{\frac{1}{s}\right\} = 1.$$

For $L^{-1}\left\{e^{-4s}\frac{1}{5}\left(\frac{(s+1)}{(s+1)^2 + 4}\right)\right\}$, we have

$$\frac{1}{5}u_4(t)e^{-(t-4)}\cos(2(t - 4)).$$

The shift of 4 units to the right is now incorporated.

For $L^{-1}\left\{e^{-4s}\frac{1}{5} \cdot \frac{1}{2}\left(\frac{2}{(s+1)^2 + 4}\right)\right\}$, we have

$$\frac{1}{10}u_4(t)e^{-(t-4)}\sin(2(t - 4)).$$

Again, the shift of 4 units to the right is now incorporated.

The general solution is the sum of the four terms:

$$y = e^{-t}\cos(2t) + \frac{1}{5}u_4(t)\left(1 - e^{-(t-4)}\cos(2(t - 4)) - \frac{1}{2}e^{-(t-4)}\sin(2(t - 4))\right).$$

The $u_4(t)$ was factored to the front of the last three terms.

To summarize,

When $t < 4$, then $u_4(t) = 0$ and we have $y = e^{-t} \cos 2t$.

When $t \geq 4$, then $u_4(t) = 1$ and we have

$$y = e^{-t} \cos 2t + \frac{1}{5} - \frac{1}{5}e^{-(t-4)} \cos(2(t-4)) - \frac{1}{10}e^{-(t-4)} \sin(2(t-4)).$$

At $t = 4$, the function is continuous

Its graph is:

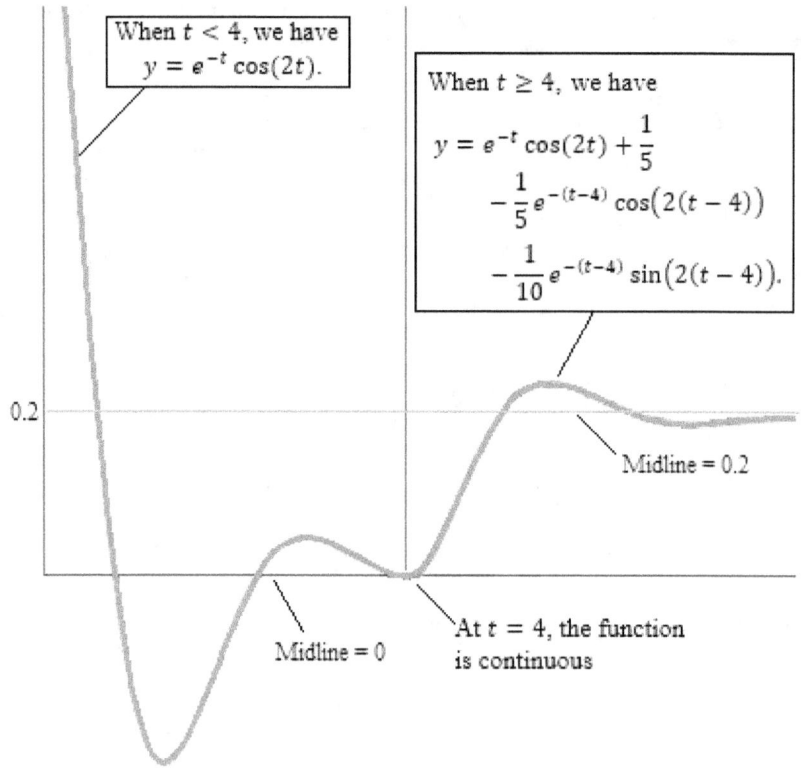

Example 21.2: Solve $y'' + 9y = \begin{cases} t, & 0 \le t < 1 \\ 2 - t, & 1 \le t \end{cases}$, $y(0) = 0$, $y'(0) = 0$.

Solution: We need to write the forcing function using $u_c(t)$ notation.

When $0 \le t < 1$, we have t, which does not need a leading "u" for now. When $1 \le t$, we need to "turn off" t and "turn on" $2 - t$. Thus, we have:

$$\underbrace{t - u_1(t) t}_{\text{This turns off } t} + \underbrace{u_1(t) (2 - t)}_{\text{This turns on } 2-t}$$

Simplify:

$$t - u_1(t)t + 2u_1(t) - u_1(t)t = t + 2u_1(t) - 2u_1(t)t$$
$$= t + u_1(t)(2 - 2t).$$

Note that when $t \ge 1$, then $u_1(t) = 1$, so that the last line is $t + 1(2 - 2t) = t + 2 - 2t$, which simplifies to $2 - t$, as in the original statement.

The IVP is now written $y'' + 9y = t + u_1(t)(2 - 2t)$, $y(0) = 0$, $y'(0) = 0$.

Apply the Laplace Transform operator to both sides:

$$L\{y''\} + 9L\{y\} = L\{t + u_1(t)(2 - 2t)\}.$$

Expand:

$$s^2 L\{y\} - sy(0) - y'(0) + 9L\{y\} = L\{t\} + L\{u_1(t)2 - u_1(t)(2(t - 1 + 1)\}.$$

Note that $y(0) = 0$ and $y'(0) = 0$:

$$s^2 L\{y\} + 9L\{y\} = L\{t\} + L\{u_1(t)2 - u_1(t)2(t - 1) - u_1(t)2\}.$$

Collect terms to each side, where $L\{y\}$ is to one side and will be isolated:

$$L\{y\}[s^2 + 9] = L\{t\} - L\{u_1(t)2(t - 1)\}.$$

Apply the Laplace Transform on the right side:

$$L\{y\}[s^2 + 9] = \frac{1 - 2e^{-s}}{s^2}.$$

Isolate $L\{y\}$:

$$L\{y\} = \frac{1}{s^2(s^2+9)} - \frac{2e^{-s}}{s^2(s^2+9)}.$$

Then the solution is

$$y = L^{-1}\left\{\frac{1}{s^2(s^2+9)} - \frac{2e^{-s}}{s^2(s^2+9)}\right\}.$$

The usual routine still applies: We need to decompose the rational expressions using partial fractions. In both cases,

$$\frac{1}{s^2(s^2+9)} = \frac{A}{s} + \frac{B}{s^2} + \frac{Cs+D}{s^2+9} = \frac{As(s^2+9) + B(s^2+9) + (Cs+D)s^2}{s^2(s^2+9)}.$$

Using methods shown in previous examples, we find that $A = 0$, $C = 0$, $B = 1/9$ and $D = -1/9$. Thus,

$$\frac{1}{s^2(s^2+9)} = \frac{1}{9}\left(\frac{1}{s^2} - \frac{1}{s^2+9}\right).$$

When $0 \le t < 1$, we have

$$y = L^{-1}\left\{\frac{1}{s^2(s^2+9)}\right\} = \frac{1}{9}L^{-1}\left\{\frac{1}{s^2}\right\} - \frac{1}{9}\cdot\frac{1}{3}L^{-1}\left\{\frac{3}{s^2+9}\right\}$$

$$= \frac{1}{9}t - \frac{1}{27}\sin 3t, \quad \text{where } 0 \le t < 1.$$

For $1 \le t$, we find $L^{-1}\left\{\frac{2e^{-s}}{s^2(s^2+9)}\right\}$. The expression $\frac{2}{s^2(s^2+9)}$ decomposes as

$$\frac{2}{s^2(s^2+9)} = \frac{2}{9}\left(\frac{1}{s^2} - \frac{1}{s^2+9}\right).$$

Thus,

$$y = L^{-1}\left\{\frac{2e^{-s}}{s^2(s^2+9)}\right\} = \frac{2}{9}u_1(t)\left[L^{-1}\left\{\frac{1}{s^2}\right\} - \frac{1}{3}\cdot L^{-1}\left\{\frac{3}{s^2+9}\right\}\right].$$

This gives

$$y = \frac{2}{9}u_1(t)\left((t-1) - \frac{1}{3}\sin(3(t-1))\right), \quad \text{where} \quad 1 \le t < 2.$$

Note that the shift of 1 unit to the right is now incorporated.

Combining all terms, the solution of $y'' + 9y = t + u_1(t)(2-2t)$, where $y(0) = 0$, $y'(0) = 0$, is

$$y = \frac{1}{9}t - \frac{1}{27}\sin 3t + \frac{2}{9}u_1(t)\left(t - 1 - \frac{1}{3}\sin(3t-3)\right).$$

Section 22
Impulse Forcing Functions

Sometimes, energy is contributed into a system instantaneously. These are called **impulses**.

Consider the function $y = f(t) = \begin{cases} 1/(2k), & -k \le t \le k \\ 0, & t < -k \text{ or } t > k \end{cases}$.

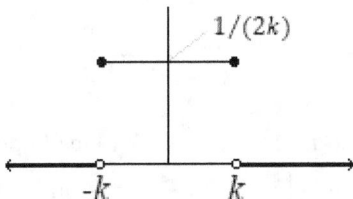

The main thing to observe here is that the area under the horizontal bar is 1.

The parameter k can be allowed to trend to 0 as a limit. However, we require that the area below the horizontal bar remain 1 unit. As $|k| \to 0$, the height of the bar trends to infinity. The function is non-zero for increasing smaller amounts of time. In this way, the **unit impulse function** δ is defined as

$$\delta(t) = 0 \text{ for all } t \ne 0, \quad \text{and} \quad \int_{-\infty}^{\infty} \delta(t)\,dt = 1.$$

The impulse can occur anywhere. If it occurs at $t = c$, we build in the shift, $\delta(t - c) = 0$ for all $t \ne c$. The integral expression (governing the area) stays the same. This function is also known as the Dirac Delta Function.

The Laplace Transform of $f(t) = \delta(t-c)$ is

$$H(s) = L\{\delta(t-c)\} = e^{-cs}.$$

Example 22.1: Find the solution of $y'' + 9y = \delta(t-4)$, with initial conditions $y(0) = 0$ and $y'(0) = 0$.

Solution: Apply the Laplace Transform operator to both sides:

$$L\{y'' + 9y\} = L\{\delta(t-4)\}$$
$$L\{y''\} + 9L\{y\} = L\{\delta(t-4)\}$$
$$s^2 L\{y\} - s\underbrace{y(0)}_{0} - \underbrace{y'(0)}_{0} + 9L\{y\} = e^{-4s}$$
$$L\{y\}[s^2 + 9] = e^{-4s}$$

Thus,

$$L\{y\} = \frac{e^{-4s}}{s^2 + 9}.$$

The solution is

$$y = L^{-1}\left\{\frac{e^{-4s}}{s^2 + 9}\right\}.$$

The e^{-4s} inverts back to $u_4(t)$ ... not back to the impulse function. The $\frac{1}{s^2+9}$ inverts as follows: $L^{-1}\left\{\frac{1}{s^2+9}\right\} = \frac{1}{3}L^{-1}\left\{\frac{3}{s^2+9}\right\} = \frac{1}{3}\sin 3t$. Thus, the solution is

$$y = \frac{1}{3}u_4(t)\sin(3(t-4)).$$

When $t < 4$, the factor $u_4(t) = 0$, so $y = 0$. Then at $t = 4$, an impulse of energy is instantaneously applied to the system, setting in motion the solution, $y = \frac{1}{3}\sin(3t - 12)$, which starts at $t = 4$ and continues forever as $t \geq 4$.

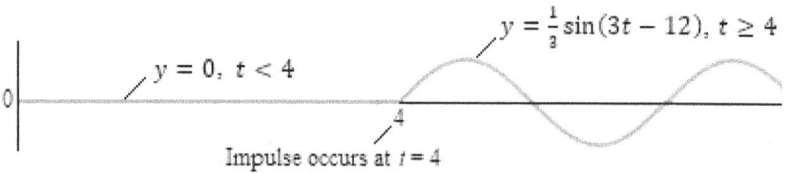

Impulse occurs at $t = 4$

Example 22.2: Solve $y'' + 5y' + 6y = \delta(t-3) + \delta(t-6)$, where $y(0) = 0$ and $y'(0) = 0$.

Solution: Appy the Laplace Transform operator to both sides:

$$L\{y''\} + 5L\{y'\} + 6L\{y\} = L\{\delta(t-3)\} + L\{\delta(t-6)\}.$$

Now simplify and isolate $L\{y\}$:

$$s^2 L\{y\} - s\underbrace{y(0)}_{0} - \underbrace{y'(0)}_{0} + 5\left(sL\{y\} - \underbrace{y(0)}_{0}\right) + 6L\{y\} = e^{-3s} + e^{-6s}$$

$$L\{y\}[s^2 + 5s + 6] = e^{-3s} + e^{-6s}$$

$$L\{y\} = \frac{e^{-3s} + e^{-6s}}{s^2 + 5s + 6}.$$

The solution is

$$y = L^{-1}\left\{\frac{e^{-3s} + e^{-6s}}{s^2 + 5s + 6}\right\}.$$

Ignoring the e terms for the moment, we have

$$\frac{1}{s^2 + 5s + 6} = \frac{1}{(s+2)(s+3)} = \frac{1}{s+2} - \frac{1}{s+3}.$$

Inverting, we have

$$L^{-1}\left\{\frac{1}{s+2}\right\} = e^{-2t} \quad \text{and} \quad L^{-1}\left\{-\frac{1}{s+3}\right\} = -e^{-3t}.$$

The e^{-3s} term inverts to $u_3(t)$ and the e^{-6s} term inverts to $u_6(t)$. Shifts are also accounted for in the inversion. Thus, the solution is

$$y = u_3(t)\left(e^{-2(t-3)} - e^{-3(t-3)}\right) + u_6(t)\left(e^{-2(t-6)} - e^{-3(t-6)}\right).$$
$$= u_3(t)(e^{6-2t} - e^{9-3t}) + u_6(t)(e^{12-2t} - e^{18-3t}).$$

In the graph, there are two "impulses" at $t = 3$ and $t = 6$:

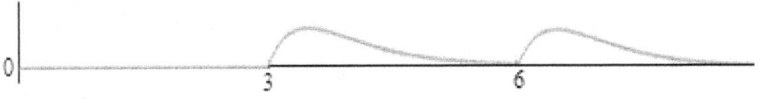

Section 23
Laplace Transforms of Special Cases

The following are examples of the Laplace Transform being applied to derivatives, shifts, horizontal stretches and the Gamma function.

Derivative Rule: If $L\{f(t)\} = H(s)$, then $L\{-tf(t)\} = H'(s)$.

Proof: Using the definition of the Laplace Transform, we have

$$L\{f(t)\} = \int_0^\infty f(t)e^{-st}\, dt.$$

Differentiate both sides with respect to s:

$$\frac{d}{ds}H(s) = \frac{d}{ds}L\{f(t)\} = \frac{d}{ds}\int_0^\infty f(t)e^{-st}\, dt.$$

The integrand can be differentiated in place with respect to s. In this step, $f(t)$ acts as a constant multiplier. Note that $\frac{d}{ds}e^{-st} = -te^{-st}$:

$$\frac{d}{ds}\int_0^\infty f(t)e^{-st}\, dt = \int_0^\infty \frac{d}{ds}(f(t)e^{-st})\, dt = \int_0^\infty -te^{-st}f(t)\, dt.$$

Thus,

$$H'(s) = \int_0^\infty -te^{-st}f(t)\, dt = L\{-tf(t)\}, \text{ where } L\{f(t)\} = H(s).$$

Corollary:

$$H''(s) = L\{(-t)(-t)f(t)\} = L\{t^2 f(t)\}.$$

In general,

$$H^{(n)}(s) = L\{(-t)^n f(t)\}.$$

This allows us to handle cases where a function may be preceded by a power of t. It allows us to handle cases of the form $L\{t^n f(t)\}$

Example 23.1: Find $L\{t\cos t\}$.

Solution: First, find $L\{\cos t\}$, which is

$$H(s) = L\{\cos t\} = \frac{s}{s^2+1}.$$

Differentiating H with respect to s, we have

$$H'(s) = \frac{1-s^2}{(s^2+1)^2}.$$

Since $L\{-tf(t)\} = H'(s)$, then moving the negative, we have

$$L\{tf(t)\} = -H'(s).$$

Therefore,

$$L\{t\cos t\} = -\left(\frac{1-s^2}{(s^2+1)^2}\right) = \frac{s^2-1}{(s^2+1)^2}.$$

Example 23.2: Find $L\{te^{2t}\}$ and $L\{t^2 e^{2t}\}$.

Solution: Start with $H(s) = L\{e^{2t}\} = \frac{1}{s-2}$.

Differentiate twice, so that

$$H'(s) = -\frac{1}{(s-2)^2} \quad \text{and} \quad H''(s) = \frac{2}{(s-2)^3}.$$

We have $L\{-te^{2t}\} = H'(s)$ or equivalently,

$$L\{te^{2t}\} = -H'(s) = -\left(-\frac{1}{(s-2)^2}\right) = \frac{1}{(s-2)^2}.$$

Similarly, we have

$$L\{(-t)^2 e^{2t}\} = L\{t^2 e^{2t}\} = H''(s) = \frac{2}{(s-2)^3}.$$

If $H(s)$ appears to have the form of a derivative, then this process can be reversed in the inversion step.

Example 23.3: Find $y = L^{-1}\left\{\frac{4}{(s+1)^3}\right\}$.

Solution: The function $H(s) = \frac{4}{(s+1)^3}$ appears to be in the form of a derivative. Integrating twice, we get

$$\int \frac{4}{(s+1)^3}\,ds = -\frac{2}{(s+1)^2} \quad \text{and} \quad \int \frac{-2}{(s+1)^2}\,ds = \frac{2}{s+1}.$$

Thus,

$$L^{-1}\left\{\frac{2}{s+1}\right\} = 2e^{-t}$$

Since $L\{t^2 f(t)\} = H''(s)$, we have

$$y = L^{-1}\left\{\frac{4}{(s+1)^3}\right\} = 2t^2 e^{-t}.$$

Comment: the function $H(s) = \frac{4}{(s+1)^3}$ is close in form to $H(s) = \frac{4}{s^3}$, in which case the inversion follows the form $L\{t^n\} = \frac{n!}{s^{n+1}}$ with the 4 in the numerator having no effect other than as a coefficient. We will revisit this same problem soon in discussing the shift rule.

Example 23.4: Use Laplace Transforms to solve $y'' - 6y' + 9y = 0$, with initial conditions $y(0) = 1$, $y'(0) = 0$.

Solution: We have

$$L\{y''\} - 6L\{y'\} + 9L\{y\} = L\{0\}$$

$$s^2 L\{y\} - s\underbrace{y(0)}_{1} - \underbrace{y'(0)}_{0} - 6\left(sL\{y\} - \underbrace{y(0)}_{1}\right) + 9L\{y\} = 0$$

$$s^2 L\{y\} - s - 6sL\{y\} + 6 + 9L\{y\} = 0$$

$$L\{y\}[s^2 - 6y + 9] = s - 6$$

$$L\{y\} = \frac{s-6}{s^2 - 6s + 9}.$$

Thus,
$$y = L^{-1}\left\{\frac{s-6}{s^2 - 6s + 9}\right\}.$$

Using partial fractions, we have
$$\frac{s-6}{s^2 - 6s + 9} = \frac{A}{s-3} + \frac{B}{(s-3)^2}.$$

In the second term, the denominator has a linear factor multiplicity 2.

Solving for A and B, we get
$$\frac{s-6}{s^2 - 6s + 9} = \frac{1}{s-3} - \frac{3}{(s-3)^2}.$$

The solution is
$$y = L^{-1}\left\{\frac{1}{s-3} - \frac{3}{(s-3)^2}\right\} = L^{-1}\left\{\frac{1}{s-3}\right\} + L^{-1}\left\{\frac{-3}{(s-3)^2}\right\}.$$

The first term is inverted and is
$$L^{-1}\left\{\frac{1}{s-3}\right\} = e^{3t}.$$

For the second term, recognize that $\frac{-3}{(s-3)^2}$ is the derivative of $\frac{3}{s-3}$.

Using the rule $L\{-te^{2t}\} = H'(s)$, we have
$$L^{-1}\left\{\frac{3}{s-3}\right\} = 3e^{3t} \quad \text{so that} \quad L^{-1}\left\{\frac{-3}{(s-3)^2}\right\} = -3te^{3t}.$$

Remember to attach a negative because of the leading negative in $L\{-te^{2t}\}$.

The solution of $y'' - 6y' + 9y = 0$, $y(0) = 1$, $y'(0) = 0$ is
$$y = e^{3t} - 3te^{3t}.$$

Shift Rule: If $L\{f(t)\} = H(s)$, then $L\{e^{at}f(t)\} = H(s-a)$.

Proof: We have $L\{f(t)\} = \int_0^\infty f(t)e^{-st}\,dt$, so that

$$L\{e^{at}f(t)\} = \int_0^\infty e^{at}f(t)e^{-st}\,dt$$

$$= \int_0^\infty f(t)e^{(a-s)t}\,dt$$

$$= \int_0^\infty f(t)e^{-(s-a)t}\,dt$$

$$= H(s-a).$$

Example 23.5: Find $L\{te^{2t}\}$ and $L\{t^2e^{2t}\}$.

Solution: Start with $L\{t\} = \frac{1}{s^2}$ and $L\{t^2\} = \frac{2}{s^3}$.

The presence of e^{2t} suggest to shift both results by 2 units. Thus,

$$L\{te^{2t}\} = \frac{1}{(s-2)^2} \quad \text{and} \quad L\{t^2e^{2t}\} = \frac{2}{(s-2)^3}.$$

This is a repeat of Example 23.2. Both methods work.

Previously we discussed the case of $H(s) = L\{t^n\}$ when n is a positive integer. Now we explore the cases for all Real numbers $n > -1$, which allows us to find Laplace Transforms of square root functions and so forth.

The Gamma Function: If $y = t^n$, where $n > -1$, then $L\{t^n\} = \frac{\Gamma(n+1)}{s^{n+1}}$, where Γ represents the gamma function.

Proof: Using the Laplace Transform, we have $L\{t^n\} = \int_0^\infty t^n e^{-st}\,dt$. Using integration by parts where $u = t^n$ and $dv = e^{-st}$, we have $du = nt^{n-1}$ and $v = -\frac{1}{s}e^{-st}$. Thus,

$$L\{t^n\} = \int_0^\infty t^n e^{-st}\,dt = \left[-\frac{1}{s}t^n e^{-st}\right]_0^\infty - \int_0^\infty -\frac{1}{s}e^{-st}(nt^{n-1})\,dt.$$

The term $\left[-\frac{1}{s}t^n e^{-st}\right]_0^\infty = 0$ after evaluation using limits.

Simplified, we have

$$\int_0^\infty t^n e^{-st} \, dt = \frac{n}{s}\int_0^\infty t^{n-1}e^{-st}\, dt.$$

This is the same as

$$L\{t^n\} = \frac{n}{s}L\{t^{n-1}\}.$$

This can be iterated:

$$L\{t^n\} = \frac{n}{s}L\{t^{n-1}\} = \frac{n}{s}\left(\frac{n-1}{s}L\{t^{n-2}\}\right) = \frac{n(n-1)}{s^2}L\{t^{n-2}\},$$

and so on.

If n is an integer such that $n \geq 0$, then this formula is the same as $L\{t^n\} = \frac{n!}{s^{n+1}}$ (Try it for $n = 3$, for example).

If n is a non-integer such that $n > -1$, then the numerator is given by the gamma function,

$$\Gamma(n+1) = \int_0^\infty t^n e^{-t}\, dt.$$

The gamma function has the property that for $n > -1$, $\Gamma(n+1) = n!$, and it "extends" the notion of factorial to include non-integers greater than -1. The integral $\int_0^\infty t^n e^{-t}\, dt$ is usually evaluated using numerical methods such a calculator.

Example 23.6: Find $L\{\sqrt{t}\}$.

Solution: We have

$$L\{\sqrt{t}\} = L\{t^{1/2}\} = \frac{\Gamma(3/2)}{s^{1/2+1}} = \frac{\Gamma(3/2)}{s^{3/2}} \approx \frac{0.88623}{s^{3/2}}.$$

An interesting corollary of this example is the fact that $\Gamma(3/2) = (1/2)!$. This number is $\sqrt{\pi}/2$. Try it on a calculator.

The $f(ct)$ Rule: If $H(s) = L\{f(t)\}$, then $L\{f(ct)\} = \frac{1}{c}H\left(\frac{s}{c}\right)$.

Proof: Start with $L\{f(ct)\} = \int_0^\infty f(ct)e^{-st}\,dt$. Let $u = ct$ so that $t = \frac{u}{c}$ and $dt = \frac{du}{c}$. Make the substitutions:

$$L\{f(ct)\} = \int_0^\infty f(ct)e^{-st}\,dt$$

$$= \int_0^\infty f(u)e^{-s\left(\frac{u}{c}\right)}\frac{du}{c}$$

$$= \frac{1}{c}\int_0^\infty f(u)e^{\left(-\frac{s}{c}\right)u}\,du$$

$$= \frac{1}{c}H\left(\frac{s}{c}\right).$$

Example 23.7: Find $L\{\sin(2t)\}$.

Solution: Starting with $L\{\sin(t)\} = \frac{1}{s^2+1}$ and observing that $c = 2$, we have

$$L\{\sin(2t)\} = \frac{1}{2}\cdot\frac{1}{\left(\frac{s}{2}\right)^2+1} = \frac{1}{2}\cdot\frac{1}{\left(\frac{s^2+4}{4}\right)} = \frac{1}{2}\cdot\left(\frac{4}{s^2+4}\right) = \frac{2}{s^2+4}.$$

This form was already known from earlier discussion, but this shows there is more than one way to find the Laplace Transform of a given function.

Section 24
Laplace Transforms of Periodic Functions

A function f is **periodic** with period T if $f(t+T) = f(t)$ for all t, where T is the smallest non-zero value for which $f(t+T) = f(t)$. For example, $f(t) = \cos t$ is periodic with periods $n\pi$, where n is an even integer. The smallest such period is 2π.

Assume $y = f(t)$ is periodic with period T. Apply the Laplace Transform operator:

$$L\{f(t)\} = \int_0^\infty f(t)e^{-st}\,dt.$$

Write the integral as a sum of two integrals, from $0 \leq t < T$ and $T \leq t < \infty$:

$$\int_0^\infty f(t)e^{-st}\, dt = \int_0^T f(t)e^{-st}\, dt + \int_T^\infty f(t+T)e^{-st}\, dt.$$

In the second integral, substitute $u = t + T$, so that $t = u - T$. Note that $du = dt$ (since T is a constant) and that with this shift, the integral in u has a lower bound of 0:

$$\int_0^\infty f(t)e^{-st}\, dt = \int_0^T f(t)e^{-st}\, dt + \int_0^\infty f(u)e^{-s(u-T)}\, du.$$

Observe that $e^{-s(u-T)} = e^{-su}e^{-sT}$. Bring the e^{-sT} outside as it is constant with respect to u:

$$\int_0^\infty f(t)e^{-st}\, dt = \int_0^T f(t)e^{-st}\, dt + e^{-sT}\int_0^\infty f(u)e^{-su}\, du.$$

Both $\int_0^\infty f(t)e^{-st}\, dt$ and $\int_0^\infty f(u)e^{-su}\, du$ are identical integrals since the variables of integration are dummy variables.

Replace the first and last integrals with $L\{f(t)\}$:

$$L\{f(t)\} = \left(\int_0^T f(t)e^{-st}\, dt\right) + e^{-sT}L\{f(t)\}.$$

Solve for $L\{f(t)\}$:

$$L\{f(t)\} - e^{-sT}L\{f(t)\} = \int_0^T f(t)e^{-st}\, dt$$

$$L\{f(t)\}(1 - e^{-sT}) = \int_0^T f(t)e^{-st}\, dt$$

$$L\{f(t)\} = \frac{\int_0^T f(t)e^{-st}\, dt}{1 - e^{-sT}}.$$

This is the Laplace Transform for a function with period T.

Example 24.1: Find $L\{\sin(t)\}$ using the formula $L\{f(t)\} = \dfrac{\int_0^T f(t)e^{-st}\,dt}{1-e^{-sT}}$.

Solution: The sine function has period $T = 2\pi$, so we have

$$L\{\sin(t)\} = \frac{\int_0^{2\pi} \sin(t)\,e^{-st}\,dt}{1 - e^{-2\pi s}}.$$

The integral $\int_0^{2\pi} \sin(t)\,e^{-st}\,dt$ is evaluated using integration-by-parts. We have

$$\int_0^{2\pi} \sin(t)\,e^{-st}\,dt = \left[-\frac{1}{s}\sin(t)\,e^{-st}\right]_0^{2\pi} + \frac{1}{s}\int_0^{2\pi} \cos(t)\,e^{-st}\,dt.$$

The term $\left[-\dfrac{1}{s}\sin(t)\,e^{-st}\right]_0^{2\pi} = 0$ after evaluation at the bounds so it vanishes.

So we have

$$\int_0^{2\pi} \sin(t)\,e^{-st}\,dt = \frac{1}{s}\int_0^{2\pi} \cos(t)\,e^{-st}\,dt.$$

Integrate by parts again on the right-most integral:

$$\int_0^{2\pi} \cos(t)\,e^{-st}\,dt = \left[-\frac{1}{s}e^{-st}\cos(t)\right]_0^{2\pi} - \frac{1}{s}\int_0^{2\pi} \sin(t)\,e^{-st}\,dt.$$

The term $\left[-\dfrac{1}{s}e^{-st}\cos(t)\right]_0^{2\pi} = -\dfrac{1}{s}e^{-2\pi s} + \dfrac{1}{s}$ after evaluating the bounds.

Making substitutions, we have

$$\int_0^{2\pi} \sin(t)\,e^{-st}\,dt = \frac{1}{s}\left[\left(-\frac{1}{s}e^{-2\pi s} + \frac{1}{s}\right) - \frac{1}{s}\int_0^{2\pi} \sin(t)\,e^{-st}\,dt\right].$$

Distribute:

$$\int_0^{2\pi} \sin(t)\,e^{-st}\,dt = \left(-\frac{1}{s^2}e^{-2\pi s} + \frac{1}{s^2}\right) - \frac{1}{s^2}\int_0^{2\pi} \sin(t)\,e^{-st}\,dt.$$

Collect the definite integrals to the left side:

$$\int_0^{2\pi} \sin(t)\,e^{-st}\,dt + \frac{1}{s^2}\int_0^{2\pi} \sin(t)\,e^{-st}\,dt = \frac{1 - e^{-2\pi s}}{s^2}.$$

Combine the two definite integrals into a single integral, where $1 + \frac{1}{s^2} = \frac{s^2+1}{s^2}$:

$$\left(\frac{s^2+1}{s^2}\right) \int_0^{2\pi} \sin(t) e^{-st} \, dt = \frac{1 - e^{-2\pi s}}{s^2}.$$

Solving for the integral, we have

$$\int_0^{2\pi} \sin(t) e^{-st} \, dt = \frac{1 - e^{-2\pi s}}{s^2} \cdot \frac{s^2}{s^2+1} = \frac{1 - e^{-2\pi s}}{s^2+1}.$$

Recall that

$$L\{\sin t\} = \frac{\int_0^{2\pi} \sin(t) e^{-st} \, dt}{1 - e^{-2\pi s}}.$$

Make the substitution in the numerator, and simplify:

$$L\{\sin t\} = \frac{1}{1 - e^{-2\pi s}} \left(\frac{1 - e^{-2\pi s}}{s^2+1}\right) = \frac{1}{s^2+1}.$$

Example 24.2: Find the Laplace Transform of the sawtooth function given by $f(t) = t$ for $0 \le t < 1$ and $f(t - 1) = f(t)$ for $t \ge 1$.

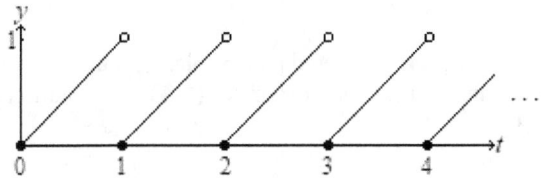

Solution: The period is $T = 1$, so we have

$$L\{f(t)\} = \frac{\int_0^1 t e^{-st} \, dt}{1 - e^{-s(1)}}.$$

The integral is evaluated using integration-by-parts:

$$\int_0^1 t e^{-st} \, dt = \left[-\frac{t}{s} e^{-st}\right]_0^1 + \frac{1}{s} \int_0^1 e^{-st} \, dt$$

$$= -\frac{1}{s} e^{-s} - \frac{1}{s^2} e^{-s} + \frac{1}{s^2} = \frac{1 - e^{-s} - se^{-s}}{s^2}.$$

Thus,

$$L\{f(t)\} = \frac{\int_0^1 te^{-st}\,dt}{1-e^{-s(1)}}$$

$$= \frac{1-e^{-s}-se^{-s}}{s^2} \cdot \frac{1}{1-e^{-s}}$$

$$= \frac{1-e^{-s}(s+1)}{s^2(1-e^{-s})}.$$

Section 25
Matrix Review:
Determinants, Eigenvalues and Eigenvectors

Matrices are commonly used to solve systems of differential equations. Often, a nth ordered linear differential equation can be rewritten as a first-ordered linear differential equation in matrix form.

We review some common aspects of matrices:

A **linear system** is two or more linear equations in two or more variables taken together. For example, $\begin{array}{l} 3x+2y=11 \\ -x+5y=19 \end{array}$ is a system of two linear equations in two variables.

A **solution of a system** is any ordered pair (triple, *etc.*) that solves all equations of the system simultaneously. For example, (1,4) is a solution of the above system, since $3(1)+2(4)=3+8=11$ is true, and $-(1)+5(4)=-1+20=19$ is also true.

A system is **consistent** if it has at least one solution. Otherwise, the system is **inconsistent**. The above system is consistent. Linear systems may have no solution (inconsistent), one solution or infinitely many solutions (both consistent).

Matrices (singular: Matrix)

Matrices are used to solve systems of linear equations as well as to give insight as to the structure of the system. There are many ways to solve a linear system using matrices. Gaussian Row Operations are one common method, Cramer's Rule is also commonly used. The system $\begin{array}{l} 3x+2y=11 \\ -x+5y=19 \end{array}$ can be written as an equation of matrices (following page):

$$\underbrace{\begin{bmatrix} 3 & 2 \\ -1 & 5 \end{bmatrix}}_{A} \underbrace{\begin{bmatrix} x \\ y \end{bmatrix}}_{X} = \underbrace{\begin{bmatrix} 11 \\ 19 \end{bmatrix}}_{B}.$$

Matrix A is the coefficient matrix, X is the variable matrix, and B is the constants matrix. Since the system has two equations in two variables, A is a 2×2 matrix, or a **square** matrix. Matrices X and B are called **vectors** since they each have just one column.

All square matrices have associated with each a unique real number called its **determinant**. For the 2×2 case, the formula is:

$$\det \begin{bmatrix} a & b \\ c & d \end{bmatrix} = ad - bc.$$

Here, "det" stands for determinant. It acts like a function that assigns to each square matrix a number, where that number is given by the above formula.

Examples:
$$\det \begin{bmatrix} 2 & 6 \\ 3 & 7 \end{bmatrix} = (2)(7) - (3)(6) = 14 - 18 = -4.$$

$$\det \begin{bmatrix} 4 & 2 \\ -5 & 6 \end{bmatrix} = (4)(6) - (-5)(2) = 24 - (-10) = 34.$$

$$\det \begin{bmatrix} 3 & 6 \\ 12 & 24 \end{bmatrix} = (3)(24) - (12)(6) = 72 - 72 = 0.$$

If the determinant of a square matrix is 0, that matrix is called **singular**.

The system $\begin{matrix} 3x + 2y = 11 \\ -x + 5y = 19 \end{matrix}$ can be written as $\underbrace{\begin{bmatrix} 3 & 2 \\ -1 & 5 \end{bmatrix}}_{A} \underbrace{\begin{bmatrix} x \\ y \end{bmatrix}}_{X} = \underbrace{\begin{bmatrix} 11 \\ 19 \end{bmatrix}}_{B}.$

Matrices X and B are both 2×1 in size (2 rows, 1 column). **Matrix multiplication** is defined as the linear combinations of rows of the left factor with columns of the right factor. Thus, if $x = 1$ and $y = 4$, we have

$$\begin{bmatrix} 3 & 2 \\ -1 & 5 \end{bmatrix} \begin{bmatrix} 1 \\ 4 \end{bmatrix} = \begin{bmatrix} 3 \cdot 1 + 2 \cdot 4 \\ -1 \cdot 1 + 5 \cdot 4 \end{bmatrix} = \begin{bmatrix} 11 \\ 19 \end{bmatrix}.$$

Matrices can also be multiplied by a constant, called a **scalar**. For example:

$$2A = 2 \begin{bmatrix} 3 & 2 \\ -1 & 5 \end{bmatrix} = \begin{bmatrix} 2 \cdot 3 & 2 \cdot 2 \\ 2 \cdot (-1) & 2 \cdot 5 \end{bmatrix} = \begin{bmatrix} 6 & 4 \\ -2 & 10 \end{bmatrix}.$$

$$7B = 7 \begin{bmatrix} 11 \\ 19 \end{bmatrix} = \begin{bmatrix} 7 \cdot 11 \\ 7 \cdot 19 \end{bmatrix} = \begin{bmatrix} 77 \\ 133 \end{bmatrix}.$$

Eigenvalues and Eigenvectors

Something interesting happens with square matrices and specially-chosen scalar multiples and vectors. For example, let $A = \begin{bmatrix} 3 & 1 \\ 3 & 5 \end{bmatrix}$, and let vectors $v_1 = \begin{bmatrix} 1 \\ -1 \end{bmatrix}$ and $v_2 = \begin{bmatrix} 1 \\ 3 \end{bmatrix}$. Note what happens when we calculate Av_1 and Av_2:

$$Av_1 = \begin{bmatrix} 3 & 1 \\ 3 & 5 \end{bmatrix}\begin{bmatrix} 1 \\ -1 \end{bmatrix} = \begin{bmatrix} 2 \\ -2 \end{bmatrix}, \text{ which is the same as } 2v_1 = 2\begin{bmatrix} 1 \\ -1 \end{bmatrix} = \begin{bmatrix} 2 \\ -2 \end{bmatrix}.$$

$$Av_2 = \begin{bmatrix} 3 & 1 \\ 3 & 5 \end{bmatrix}\begin{bmatrix} 1 \\ 3 \end{bmatrix} = \begin{bmatrix} 6 \\ 18 \end{bmatrix}, \text{ which is the same as } 6v_2 = 6\begin{bmatrix} 1 \\ 3 \end{bmatrix} = \begin{bmatrix} 6 \\ 18 \end{bmatrix}.$$

For specially chosen vectors, the action of multiplying it by a matrix A has the same result as if that vector were multiplied by some non-zero scalar constant.

This doesn't always work. For example, let $v_3 = \begin{bmatrix} 2 \\ 1 \end{bmatrix}$. Then $\begin{bmatrix} 3 & 1 \\ 3 & 5 \end{bmatrix}\begin{bmatrix} 2 \\ 1 \end{bmatrix} = \begin{bmatrix} 7 \\ 11 \end{bmatrix}$. But $\begin{bmatrix} 7 \\ 11 \end{bmatrix}$ is not a scalar multiple of $v_3 = \begin{bmatrix} 2 \\ 1 \end{bmatrix}$.

In the above example cases, we call 2 and 6 the **eigenvalues** of A, and call the vectors v_1 and v_2 the **eigenvectors** of A. Eigenvalues are usually denoted by the Greek letter lambda, λ.

Finding Eigenvalues and Eigenvectors

Let A be a square matrix (we'll look at the 2×2 case for now). We seek to find a scalar eigenvalue λ and a non-zero eigenvector v such that

$$Av = \lambda v.$$

On the left side, Av represents a 2×2 matrix multiplying a 2×1 matrix. On the right side, λ is a scalar multiplied to a 2×1 matrix. To make both sides structurally the same, we rewrite λ as $\lambda I = \begin{bmatrix} \lambda & 0 \\ 0 & \lambda \end{bmatrix}$, where $I = \begin{bmatrix} 1 & 0 \\ 0 & 1 \end{bmatrix}$ is the multiplicative identity matrix. Now, both sides are a 2×2 matrix multiplying to a 2×1 matrix.

This allows us to move the terms around algebraically:

$$Av = \lambda I v \quad \text{is the same as} \quad Av - \lambda I v = 0.$$

Factoring, we have $(A - \lambda I)v = 0$. Since $v \neq 0$, then this is only true when $\det(A - \lambda I) = 0$. This is the formula we use to find eigenvalues.

Example 25.1: Find the eigenvalues and eigenvectors of $A = \begin{bmatrix} 3 & 1 \\ 3 & 5 \end{bmatrix}$.

Solution: Since $\lambda I = \begin{bmatrix} \lambda & 0 \\ 0 & \lambda \end{bmatrix}$, we have

$$\det(A - \lambda I) = \det \begin{bmatrix} 3 - \lambda & 1 \\ 3 & 5 - \lambda \end{bmatrix} = (3 - \lambda)(5 - \lambda) - 3.$$

This is set to 0 and solved for λ:

$$(3 - \lambda)(5 - \lambda) - 3 = 0$$
$$\lambda^2 - 8\lambda + 15 - 3 = 0$$
$$\lambda^2 - 8\lambda + 12 = 0$$
$$(\lambda - 2)(\lambda - 6) = 0$$
$$\lambda_1 = 2 \quad \text{and} \quad \lambda_2 = 6.$$

The two eigenvalues of matrix A are $\lambda_1 = 2$ and $\lambda_2 = 6$ where the subscripts help keep track of each eigenvalue.

To find the eigenvectors, use a generic vector $v = \begin{bmatrix} a \\ b \end{bmatrix}$, and solve $(A - \lambda I)v = 0$. We'll start with $\lambda_1 = 2$:

$$\begin{bmatrix} 3 - 2 & 1 \\ 3 & 5 - 2 \end{bmatrix} \begin{bmatrix} a \\ b \end{bmatrix} = 0$$

This simplifies to

$$\begin{bmatrix} 1 & 1 \\ 3 & 3 \end{bmatrix} \begin{bmatrix} a \\ b \end{bmatrix} = 0.$$

Note that the square matrix is singular. Its determinant is 0. For a matrix of size 2 × 2, it will be singular if there is a row of 0s, or if one row is a multiple of another row. For larger square matrices, singularity happens when one row is some linear combination of two or more of the other rows in the matrix.

Singularity always happens after subtracting the diagonal elements by the eigenvalue and is a good check of your work.

The top row multiplies to $a + b = 0$. If we choose $a = 1$, then $b = -1$, so we have an eigenvector associated with eigenvalue $\lambda_1 = 2$; it is $v_1 = \begin{bmatrix} 1 \\ -1 \end{bmatrix}$. Any non-zero multiple of this vector is also an eigenvector corresponding to $\lambda_1 = 2$. The subscripts are used to track the eigenvalue and its associated eigenvector.

For the eigenvalue $\lambda_2 = 6$, its eigenvector is also found in a similar way:

$$\begin{bmatrix} 3-6 & 1 \\ 3 & 5-6 \end{bmatrix} \begin{bmatrix} a \\ b \end{bmatrix} = 0 \quad \text{which simplifies to} \quad \begin{bmatrix} -3 & 1 \\ 3 & -1 \end{bmatrix} \begin{bmatrix} a \\ b \end{bmatrix} = 0.$$

The top row multiplies to $-3a + b = 0$. If we choose $a = 1$, then $b = 3$. We have an eigenvector associated with eigenvalue $\lambda_2 = 6$; it is $v_2 = \begin{bmatrix} 1 \\ 3 \end{bmatrix}$.

Example 25.2: Find the eigenvalues and eigenvectors of $A = \begin{bmatrix} 3 & 1 \\ 0 & 1 \end{bmatrix}$.

Solution: We have $\det \begin{bmatrix} 3-\lambda & 1 \\ 0 & 1-\lambda \end{bmatrix} = 0$, which gives $(3-\lambda)(1-\lambda) = 0$. Thus, the two eigenvalues are $\lambda_1 = 3$ and $\lambda_2 = 1$. (It makes no difference the order of the subscripts.)

The eigenvector for $\lambda_1 = 3$ is $v_1 = \begin{bmatrix} a \\ b \end{bmatrix}$, where $\begin{bmatrix} 3-3 & 1 \\ 0 & 1-3 \end{bmatrix} \begin{bmatrix} a \\ b \end{bmatrix} = \begin{bmatrix} 0 \\ 0 \end{bmatrix}$.

This simplifies to $\begin{bmatrix} 0 & 1 \\ 0 & -2 \end{bmatrix} \begin{bmatrix} a \\ b \end{bmatrix} = \begin{bmatrix} 0 \\ 0 \end{bmatrix}$. Note that $\begin{bmatrix} 0 & 1 \\ 0 & -2 \end{bmatrix}$ is singular.

Multiplying the top row by the column vector $\begin{bmatrix} a \\ b \end{bmatrix}$, we have $0a + 1b = 0$, or simply $b = 0$. Note that $0a$ does not imply that $a = 0$. Thus, *any* non-zero value may be chosen for a. We'll let $a = 1$. Therefore, the eigenvector corresponding to $\lambda_1 = 3$ is $v_1 = \begin{bmatrix} 1 \\ 0 \end{bmatrix}$.

Now we find the eigenvector for $\lambda_2 = 1$.

The eigenvector for $\lambda_2 = 1$ is $v_1 = \begin{bmatrix} a \\ b \end{bmatrix}$, where $\begin{bmatrix} 3-1 & 1 \\ 0 & 1-1 \end{bmatrix} \begin{bmatrix} a \\ b \end{bmatrix} = \begin{bmatrix} 0 \\ 0 \end{bmatrix}$.

This simplifies to $\begin{bmatrix} 2 & 1 \\ 0 & 0 \end{bmatrix} \begin{bmatrix} a \\ b \end{bmatrix} = \begin{bmatrix} 0 \\ 0 \end{bmatrix}$. Again, note that $\begin{bmatrix} 2 & 1 \\ 0 & 0 \end{bmatrix}$ is singular.

Multiplying the top row by the column vector $\begin{bmatrix} a \\ b \end{bmatrix}$, we have $2a + 1b = 0$. Letting $a = 1$ forces $b = -2$. Thus, the eigenvector of $\lambda_2 = 1$ is $v_2 = \begin{bmatrix} 1 \\ -2 \end{bmatrix}$.

Sometimes, the eigenvalues are not "convenient" integers.

Example 25.3: Find the eigenvalues and eigenvectors of $A = \begin{bmatrix} 2 & 1 \\ 3 & 2 \end{bmatrix}$.

Solution: We have $\det \begin{bmatrix} 2-\lambda & 1 \\ 3 & 2-\lambda \end{bmatrix} = 0$, which gives $\lambda^2 - 4\lambda + 1 = 0$. Using the quadratic formula, we have $\lambda_1 = 2 + \sqrt{3}$ and $\lambda_2 = 2 - \sqrt{3}$.

The eigenvector corresponding to $\lambda_1 = 2 + \sqrt{3}$ is $v_1 = \begin{bmatrix} a \\ b \end{bmatrix}$, where

$$\begin{bmatrix} 2-(2+\sqrt{3}) & 1 \\ 3 & 2-(2+\sqrt{3}) \end{bmatrix} \begin{bmatrix} a \\ b \end{bmatrix} = \begin{bmatrix} 0 \\ 0 \end{bmatrix}.$$

This simplifies to $\begin{bmatrix} -\sqrt{3} & 1 \\ 3 & -\sqrt{3} \end{bmatrix} \begin{bmatrix} a \\ b \end{bmatrix} = \begin{bmatrix} 0 \\ 0 \end{bmatrix}$. Is $\begin{bmatrix} -\sqrt{3} & 1 \\ 3 & -\sqrt{3} \end{bmatrix}$ singular? Its determinant is $(-\sqrt{3})(-\sqrt{3}) - (3)(1) = 3 - 3 = 0$, so yes, it is singular.

Multiplying the top row by $\begin{bmatrix} a \\ b \end{bmatrix}$ gives $-\sqrt{3}a + b = 0$. Letting $a = 1$, then $b = \sqrt{3}$, and an eigenvector corresponding to $\lambda_1 = 2 + \sqrt{3}$ is $v_1 = \begin{bmatrix} 1 \\ \sqrt{3} \end{bmatrix}$, or any non-zero multiple thereof.

Now we find the eigenvector of $\lambda_2 = 2 - \sqrt{3}$.

It is $v_2 = \begin{bmatrix} a \\ b \end{bmatrix}$, where $\begin{bmatrix} 2-(2-\sqrt{3}) & 1 \\ 3 & 2-(2-\sqrt{3}) \end{bmatrix} \begin{bmatrix} a \\ b \end{bmatrix} = \begin{bmatrix} 0 \\ 0 \end{bmatrix}$.

This simplifies to $\begin{bmatrix} \sqrt{3} & 1 \\ 3 & \sqrt{3} \end{bmatrix} \begin{bmatrix} a \\ b \end{bmatrix} = \begin{bmatrix} 0 \\ 0 \end{bmatrix}$. The matrix is singular (you should verify this).

Multiplying the top row by $\begin{bmatrix} a \\ b \end{bmatrix}$ gives $\sqrt{3}a + b = 0$. Letting $a = 1$, then $b = -\sqrt{3}$, and an eigenvector corresponding to $\lambda_2 = 2 - \sqrt{3}$ is $v_2 = \begin{bmatrix} 1 \\ -\sqrt{3} \end{bmatrix}$, or any non-zero multiple thereof.

Eigenvalues may not be real number.

Example 25.4: Find the eigenvalues and eigenvectors of $A = \begin{bmatrix} 1 & -4 \\ 1 & 1 \end{bmatrix}$.

Solution: We have $\det \begin{bmatrix} 1-\lambda & -4 \\ 1 & 1-\lambda \end{bmatrix} = 0$, which simplifies to $\lambda^2 - 2\lambda + 5 = 0$. Using the quadratic formula, we have $\lambda_1 = 1 + 2i$ and $\lambda_2 = 1 - 2i$.

The eigenvector for $\lambda_1 = 1 + 2i$ is $v_1 = \begin{bmatrix} a \\ b \end{bmatrix}$, where

$$\begin{bmatrix} 1-(1+2i) & -4 \\ 1 & 1-(1+2i) \end{bmatrix} \begin{bmatrix} a \\ b \end{bmatrix} = \begin{bmatrix} 0 \\ 0 \end{bmatrix}.$$

This simplifies to $\begin{bmatrix} -2i & -4 \\ 1 & -2i \end{bmatrix} \begin{bmatrix} a \\ b \end{bmatrix} = \begin{bmatrix} 0 \\ 0 \end{bmatrix}$. To be sure this is correct, we check it determinant: It is $(-2i)(-2i) - (1)(-4) = 4i^2 + 4 = -4 + 4 = 0$. It is singular.

The top row multiplied by $\begin{bmatrix} a \\ b \end{bmatrix}$ gives $-2ia - 4b = 0$. If $a = 2$, then $b = -i$. Thus, an eigenvector for $\lambda_1 = 1 + 2i$ is $v_1 = \begin{bmatrix} 2 \\ -i \end{bmatrix}$, or any non-zero scalar multiple.

Now we find an eigenvector for $\lambda_2 = 1 - 2i$. It is $v_2 = \begin{bmatrix} a \\ b \end{bmatrix}$, where

$$\begin{bmatrix} 1-(1-2i) & -4 \\ 1 & 1-(1-2i) \end{bmatrix} \begin{bmatrix} a \\ b \end{bmatrix} = \begin{bmatrix} 0 \\ 0 \end{bmatrix}.$$

This simplifies to $\begin{bmatrix} 2i & -4 \\ 1 & 2i \end{bmatrix} \begin{bmatrix} a \\ b \end{bmatrix} = \begin{bmatrix} 0 \\ 0 \end{bmatrix}$. It is singular (you check this).

The top row multiplied by $\begin{bmatrix} a \\ b \end{bmatrix}$ gives $2ia - 4b = 0$. If $a = 2$, then $b = i$. Thus, an eigenvector for $\lambda_2 = 1 - 2i$ is $v_2 = \begin{bmatrix} 2 \\ i \end{bmatrix}$, or any non-zero scalar multiple.

We look now at a larger square matrix. In the following example, each eigenvalue has multiplicity 1.

Example 25.5: Find the eigenvalues and eigenvectors of $A = \begin{bmatrix} 2 & 1 & -1 \\ 0 & 1 & 2 \\ 0 & 0 & 3 \end{bmatrix}$.

Solution: We have $\det \begin{bmatrix} 2-\lambda & 1 & -1 \\ 0 & 1-\lambda & 2 \\ 0 & 0 & 3-\lambda \end{bmatrix} = 0$, which simplifies (in factored form) to $(2-\lambda)(1-\lambda)(3-\lambda) = 0$, so we have three eigenvalues: $\lambda_1 = 2$, $\lambda_2 = 1$ and $\lambda_3 = 3$.

The eigenvector corresponding to $\lambda_1 = 2$ is a vector $\begin{bmatrix} a \\ b \\ c \end{bmatrix}$ such that

$$\begin{bmatrix} 2-2 & 1 & -1 \\ 0 & 1-2 & 2 \\ 0 & 0 & 3-2 \end{bmatrix} \begin{bmatrix} a \\ b \\ c \end{bmatrix} = \begin{bmatrix} 0 \\ 0 \\ 0 \end{bmatrix}.$$

This simplifies to $\begin{bmatrix} 0 & 1 & -1 \\ 0 & -1 & 2 \\ 0 & 0 & 1 \end{bmatrix} \begin{bmatrix} a \\ b \\ c \end{bmatrix} = \begin{bmatrix} 0 \\ 0 \\ 0 \end{bmatrix}.$

There are a couple ways to infer the values of the eigenvector. Note that after multiplying the matrices together, we get

$$b - c = 0$$
$$-b + 2c = 0$$
$$c = 0.$$

Both b and $c = 0$. But no inference is made on the value of a. It can be any real number. So we choose $a = 1$ for convenience and the eigenvector corresponding to $\lambda_1 = 2$ is

$$v_1 = \begin{bmatrix} 1 \\ 0 \\ 0 \end{bmatrix}.$$

Another way that may work better in general is to find the reduced row-echelon form of matrix A. This is abbreviated RREF and most calculators have this routine programmed into its library of algorithms. For the remaining eigenvectors of this example, we'll use the RREF method to simplify matrix A.

For $\lambda_2 = 1$, we have $\begin{bmatrix} 2-1 & 1 & -1 \\ 0 & 1-1 & 2 \\ 0 & 0 & 3-1 \end{bmatrix} \begin{bmatrix} a \\ b \\ c \end{bmatrix} = \begin{bmatrix} 0 \\ 0 \\ 0 \end{bmatrix}$, which is

$$\begin{bmatrix} 1 & 1 & -1 \\ 0 & 0 & 2 \\ 0 & 0 & 1 \end{bmatrix} \begin{bmatrix} a \\ b \\ c \end{bmatrix} = \begin{bmatrix} 0 \\ 0 \\ 0 \end{bmatrix}.$$

The 3 × 3 matrix above is equivalent to $\begin{bmatrix} 1 & 1 & 0 \\ 0 & 0 & 1 \\ 0 & 0 & 0 \end{bmatrix}$ in RREF.

We have

$$\begin{bmatrix} 1 & 1 & 0 \\ 0 & 0 & 1 \\ 0 & 0 & 0 \end{bmatrix} \begin{bmatrix} a \\ b \\ c \end{bmatrix} = \begin{bmatrix} 0 \\ 0 \\ 0 \end{bmatrix}.$$

When multiplied, the top row gives $a + b = 0$ and the second row gives $c = 0$. From the top row, if we let $a = 1$, then $b = -1$.

Thus, an eigenvector corresponding to $\lambda_2 = 1$ is $v_2 = \begin{bmatrix} 1 \\ -1 \\ 0 \end{bmatrix}$.

For $\lambda_3 = 3$, the square matrix in RREF is $\begin{bmatrix} 1 & 0 & 0 \\ 0 & 1 & -1 \\ 0 & 0 & 0 \end{bmatrix} \begin{bmatrix} a \\ b \\ c \end{bmatrix} = \begin{bmatrix} 0 \\ 0 \\ 0 \end{bmatrix}$.

When multiplied, the top row gives $a = 0$ and the second row gives $b - c = 0$. From the second row, if we let $b = 1$, then $c = 1$, so an eigenvector corresponding to $\lambda_3 = 3$ is $v_3 = \begin{bmatrix} 0 \\ 1 \\ 1 \end{bmatrix}$.

In conclusion, the three eigenvalues of $A = \begin{bmatrix} 2 & 1 & -1 \\ 0 & 1 & 2 \\ 0 & 0 & 3 \end{bmatrix}$ are $\lambda_1 = 2$, $\lambda_2 = 1$ and $\lambda_3 = 3$ and the three corresponding eigenvectors are $v_1 = \begin{bmatrix} 1 \\ 0 \\ 0 \end{bmatrix}$, $v_2 = \begin{bmatrix} 1 \\ -1 \\ 0 \end{bmatrix}$ and $v_3 = \begin{bmatrix} 0 \\ 1 \\ 1 \end{bmatrix}$.

In the following example, an eigenvalue has a multiplicity greater than 1:

Example 25.6: Find the eigenvalues and eigenvectors of $A = \begin{bmatrix} 1 & 0 & 3 \\ 0 & 1 & 2 \\ 0 & 0 & 2 \end{bmatrix}$.

Solution: For the eigenvalues, we have $\det \begin{bmatrix} 1-\lambda & 0 & 3 \\ 0 & 1-\lambda & 2 \\ 0 & 0 & 2-\lambda \end{bmatrix} = 0$, which simplifies to $(2 - \lambda)(1 - \lambda)^2 = 0$, so we have two eigenvalues: $\lambda_1 = 2$ with multiplicity 1, and $\lambda_2 = 1$ with multiplicity 2.

The eigenvector corresponding to $\lambda_1 = 2$ is a vector $\begin{bmatrix} a \\ b \\ c \end{bmatrix}$ such that

$$\begin{bmatrix} 1-2 & 0 & 3 \\ 0 & 1-2 & 2 \\ 0 & 0 & 2-2 \end{bmatrix} \begin{bmatrix} a \\ b \\ c \end{bmatrix} = \begin{bmatrix} 0 \\ 0 \\ 0 \end{bmatrix}.$$

In RREF, we have $\begin{bmatrix} 1 & 0 & -3 \\ 0 & 1 & -2 \\ 0 & 0 & 0 \end{bmatrix} \begin{bmatrix} a \\ b \\ c \end{bmatrix} = \begin{bmatrix} 0 \\ 0 \\ 0 \end{bmatrix}$. When multiplied, we get two equations: $a - 3c = 0$ and $b - 2c = 0$. Thus, if $c = 1$, we get $a = 3$ and $b = 2$. This is the eigenvector corresponding to $\lambda_1 = 2$:

$$v_1 = \begin{bmatrix} 3 \\ 2 \\ 1 \end{bmatrix}.$$

The eigenvector corresponding to $\lambda_2 = 1$ is a vector $\begin{bmatrix} a \\ b \\ c \end{bmatrix}$ such that

$$\begin{bmatrix} 1-1 & 0 & 3 \\ 0 & 1-1 & 2 \\ 0 & 0 & 2-1 \end{bmatrix} \begin{bmatrix} a \\ b \\ c \end{bmatrix} = \begin{bmatrix} 0 \\ 0 \\ 0 \end{bmatrix}.$$

In RREF, we have $\begin{bmatrix} 0 & 0 & 1 \\ 0 & 0 & 0 \\ 0 & 0 & 0 \end{bmatrix} \begin{bmatrix} a \\ b \\ c \end{bmatrix} = \begin{bmatrix} 0 \\ 0 \\ 0 \end{bmatrix}$. This implies $c = 0$. But a and b can be any number, independent of one another. Since a is chosen independently of b (and vice-versa), we build an eigenvector for each case. If we let $a = 1$, then one eigenvector that corresponds to $\lambda_2 = 1$ is $v_2 = \begin{bmatrix} 1 \\ 0 \\ 0 \end{bmatrix}$. Similarly, when $b = 1$, another eigenvector that corresponds to $\lambda_2 = 1$ is $v_3 = \begin{bmatrix} 0 \\ 1 \\ 0 \end{bmatrix}$.

To summarize, this matrix has eigenvector $\lambda_1 = 2$ of multiplicity 1, and its associated eigenvector is $v_1 = \begin{bmatrix} 3 \\ 2 \\ 1 \end{bmatrix}$.

The matrix also has an eigenvector $\lambda_2 = 1$ of multiplicity 2, and it has two eigenvectors, $v_2 = \begin{bmatrix} 1 \\ 0 \\ 0 \end{bmatrix}$ and $v_3 = \begin{bmatrix} 0 \\ 1 \\ 0 \end{bmatrix}$.

Note: in the case of $\lambda_2 = 1$, we observed that a and b in the construction of the eigenvector(s) are independent of one another. We built one eigenvector for each, rather than one for both. The vector $\begin{bmatrix} 1 \\ 1 \\ 0 \end{bmatrix}$ would not work as an eigenvector, because this implies that $a = b$, when no such inference can be made.

Generally, if a real-valued eigenvalue has multiplicity 1, it will have one associated eigenvector, and if it has multiplicity 2, it will have two associated eigenvectors, and so on.

Section 26
Systems: Real Eigenvalues of Multiplicity 1

Let $x_1(t)$ and $x_2(t)$ be two functions. A system of differential equations can have the form

$$x_1'(t) = a_1 x_1(t) + b_1 x_2(t)$$
$$x_2'(t) = a_2 x_1(t) + b_2 x_2(t)$$

where a_1, b_1, a_2 and b_2 are constants. This is an example of a linear system of ODEs with constant coefficients. Written as an equation using matrices, we have

$$\underbrace{\begin{bmatrix} x_1'(t) \\ x_2'(t) \end{bmatrix}}_{\mathbf{x}'(t)} = \underbrace{\begin{bmatrix} a_1 & b_1 \\ a_2 & b_2 \end{bmatrix}}_{A} \underbrace{\begin{bmatrix} x_1(t) \\ x_2(t) \end{bmatrix}}_{\mathbf{x}(t)}.$$

Such systems are written (short-hand) as $\mathbf{x}' = A\mathbf{x}$. It is first-order, linear and homogeneous. The general solution of such a system is

$$\mathbf{x}(t) = C_1 v_1 e^{\lambda_1 t} + C_2 v_2 e^{\lambda_2 t},$$

where λ_1 and λ_2 are the eigenvalues of A, and v_1 and v_2 are their eigenvectors, respectively.

Example 26.1: Solve $y'' - 2y' - 15y = 0$ by first rewriting this second-order linear ODE as a first-order linear ODE in matrix form.

Solution: First, rename the variables: let $x_1(t) = y$ and $x_2(t) = x_1'(t) = y'$. Note that $x_2'(t) = y''$. So now we have two equations:

$$x_1'(t) = x_2(t)$$
$$x_2'(t) - 2x_2(t) - 15x_1(t) = 0$$

Isolate the two derivatives to the left side:

$$x_1'(t) = x_2(t)$$
$$x_2'(t) = 15x_1(t) + 2x_2(t)$$

By "stacking" the terms, it is easier to infer the matrix A. In matrix form, the equivalent differential equation is

$$\mathbf{x}'(t) = \begin{bmatrix} 0 & 1 \\ 15 & 2 \end{bmatrix} \mathbf{x}(t)$$

$$\begin{bmatrix} x_1'(t) \\ x_2'(t) \end{bmatrix} = \begin{bmatrix} 0 & 1 \\ 15 & 2 \end{bmatrix} \begin{bmatrix} x_1(t) \\ x_2(t) \end{bmatrix}.$$

We will use eigenvalues and eigenvectors to solve this system. First, find the eigenvalues of A.

Start with $\det \begin{bmatrix} 0-\lambda & 1 \\ 15 & 2-\lambda \end{bmatrix} = 0$, which gives $\lambda^2 - 2\lambda - 15 = 0$ after simplification. This factors as $(\lambda - 5)(\lambda + 3) = 0$. Thus, the two eigenvalues are $\lambda_1 = 5$ and $\lambda_2 = -3$. The eigenvectors are found next.

For $\lambda_1 = 5$, we have $\begin{bmatrix} -5 & 1 \\ 15 & -3 \end{bmatrix} \begin{bmatrix} a \\ b \end{bmatrix} = \begin{bmatrix} 0 \\ 0 \end{bmatrix}$. As usual, note that the matrix is singular. Multiplying the top row with the matrix $\begin{bmatrix} a \\ b \end{bmatrix}$ gives $-5a + b = 0$. Let $a = 1$, so then $b = 5$. Thus, the corresponding eigenvector is $v_1 = \begin{bmatrix} 1 \\ 5 \end{bmatrix}$. For $\lambda_2 = -3$: we have $\begin{bmatrix} 3 & 1 \\ 15 & 5 \end{bmatrix} \begin{bmatrix} a \\ b \end{bmatrix} = \begin{bmatrix} 0 \\ 0 \end{bmatrix}$. This implies that $3a + b = 0$. When $a = 1$, then $b = -3$. The corresponding eigenvector is $v_2 = \begin{bmatrix} 1 \\ -3 \end{bmatrix}$.

The general solution is written in the form $\mathbf{x}(t) = C_1 v_1 e^{\lambda_1 t} + C_2 v_2 e^{\lambda_2 t}$.

Thus, the solution of $\begin{bmatrix} x_1'(t) \\ x_2'(t) \end{bmatrix} = \begin{bmatrix} 0 & 1 \\ 15 & 2 \end{bmatrix} \begin{bmatrix} x_1(t) \\ x_2(t) \end{bmatrix}$ is

$$\mathbf{x}(t) = C_1 \begin{bmatrix} 1 \\ 5 \end{bmatrix} e^{5t} + C_2 \begin{bmatrix} 1 \\ -3 \end{bmatrix} e^{-3t}.$$

Written in matrix form, this is

$$\mathbf{x}(t) = \begin{bmatrix} x_1(t) \\ x_2(t) \end{bmatrix} = \begin{bmatrix} C_1 e^{5t} + C_2 e^{-3t} \\ 5C_1 e^{5t} - 3C_2 e^{-3t} \end{bmatrix}.$$

Recall that we defined $x_1(t) = y$ and $x_2(t) = x_1'(t) = y'$. Look carefully and you'll see that the first row is $x_1(t) = C_1 e^{5t} + C_2 e^{-3t}$ and the second row is $x_2(t) = 5C_1 e^{5t} - 3C_2 e^{-3t}$, which is the derivative of the first row.

Check: The solution of $\begin{bmatrix} x_1'(t) \\ x_2'(t) \end{bmatrix} = \begin{bmatrix} 0 & 1 \\ 15 & 2 \end{bmatrix} \begin{bmatrix} x_1(t) \\ x_2(t) \end{bmatrix}$ is

$$\mathbf{x}(t) = \begin{bmatrix} x_1(t) \\ x_2(t) \end{bmatrix} = \begin{bmatrix} C_1 e^{5t} + C_2 e^{-3t} \\ 5C_1 e^{5t} - 3C_2 e^{-3t} \end{bmatrix}.$$

The derivative of $\mathbf{x}(t)$ is $\mathbf{x}'(t) = \begin{bmatrix} 5C_1 e^{5t} - 3C_2 e^{-3t} \\ 25C_1 e^{5t} + 9C_2 e^{-3t} \end{bmatrix}.$

Thus, we have $\begin{bmatrix} 5C_1 e^{5t} - 3C_2 e^{-3t} \\ 25C_1 e^{5t} + 9C_2 e^{-3t} \end{bmatrix} = \begin{bmatrix} 0 & 1 \\ 15 & 2 \end{bmatrix} \begin{bmatrix} C_1 e^{5t} + C_2 e^{-3t} \\ 5C_1 e^{5t} - 3C_2 e^{-3t} \end{bmatrix}.$

Multiply the right side. Note that the matrix $\begin{bmatrix} C_1 e^{5t} + C_2 e^{-3t} \\ 5C_1 e^{5t} - 3C_2 e^{-3t} \end{bmatrix}$ is of size 2×1. The first row [0 1] multiplied by $\begin{bmatrix} C_1 e^{5t} + C_2 e^{-3t} \\ 5C_1 e^{5t} - 3C_2 e^{-3t} \end{bmatrix}$ gives

$$(0)(C_1 e^{5t} + C_2 e^{-3t}) + (1)(5C_1 e^{5t} - 3C_2 e^{-3t}) = 5C_1 e^{5t} - 3C_2 e^{-3t}.$$

The second row [15 2] multiplied by $\begin{bmatrix} C_1 e^{5t} + C_2 e^{-3t} \\ 5C_1 e^{5t} - 3C_2 e^{-3t} \end{bmatrix}$ gives

$$(15)(C_1 e^{5t} + C_2 e^{-3t}) + (2)(5C_1 e^{5t} - 3C_2 e^{-3t})$$
$$= 15C_1 e^{5t} + 15C_2 e^{-3t} + 10C_1 e^{5t} - 6C_2 e^{-3t}.$$

This simplifies to $25C_1 e^{5t} + 9C_2 e^{-3t}$.

This expression and the one three lines above it are the elements of $\mathbf{x}'(t)$ from the previous page. Thus, the solution fully written out is

$$\mathbf{x}(t) = \begin{bmatrix} x_1(t) \\ x_2(t) \end{bmatrix} = C_1 \begin{bmatrix} 1 \\ 5 \end{bmatrix} e^{5t} + C_2 \begin{bmatrix} 1 \\ -3 \end{bmatrix} e^{-3t} = \begin{bmatrix} C_1 e^{5t} + C_2 e^{-3t} \\ 5C_1 e^{5t} - 3C_2 e^{-3t} \end{bmatrix}.$$

Phase Portraits

Phase portraits are a visual way to see the global behavior of solution curves for a system. A point $\mathbf{x}$ (actually, a vector, but treated as an ordered pair) is chosen in the xy-plane, and evaluated $A\mathbf{x}$, and this is then sketched as a small arrow indicating a "flow" (tangent arrow) of a solution curve through the chosen point.

Each eigenvalue and its associated eigenvector govern how the solution curves will flow. Note that is all systems of the form $\mathbf{x}' = A\mathbf{x}$ will always have the origin as a solution. This is called an **equilibrium solution** (or a node). The eigenvector forms a basis for a line (called an eigenspace). Sketch the vector and imagine a line passing through it in both directions. If the associated eigenvalue is positive, curves trend away from the origin, and if the associated eigenvalue is negative, curves trend in toward the origin. However, depending on whether both eigenvalues are positive, negative, or of different sign, the nature of the curves will vary.

The origin is stable if most curves flow in toward the origin (both eigenvalues would be negative). It is unstable if most curves flow away (both eigenvalues would be positive). If the eigenvalues are of different sign, as in the diagram that follows, all curves flow away from the origin (some will come in at first, then turn and flow away). However, any that start on the eigenspace associated with the negative eigenvalue will flow into the origin. The origin is called a saddle.

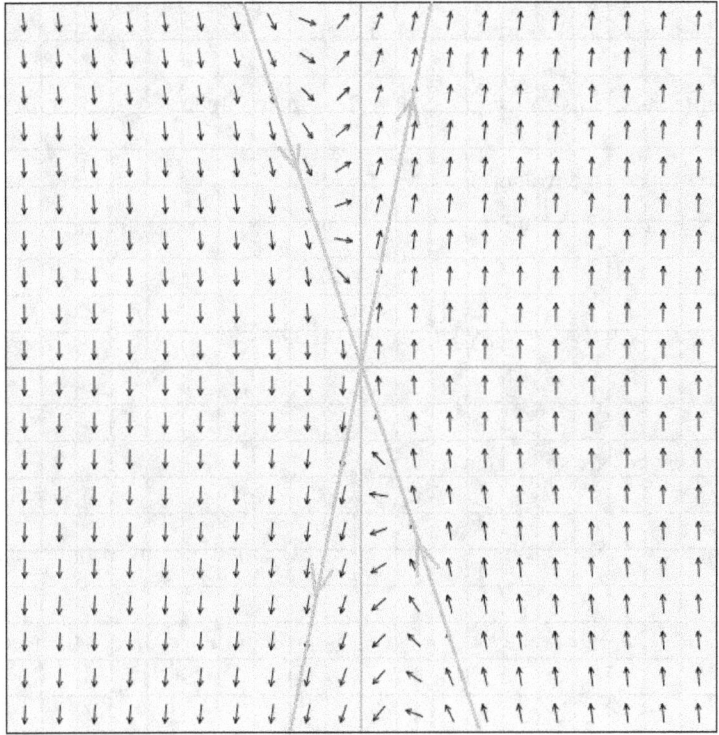

Scale: each gridline is 0.5 unit. Courtesy https://aeb019.hosted.uark.edu/pplane.html.

Above is the phase diagram of the system

$$\mathbf{x}'(t) = \begin{bmatrix} 0 & 1 \\ 15 & 2 \end{bmatrix} \mathbf{x}(t)$$ in Example 26.1.

Note the eigenvectors $v_1 = \begin{bmatrix} 1 \\ 5 \end{bmatrix}$ and $v_2 = \begin{bmatrix} 1 \\ -3 \end{bmatrix}$ each form a basis for individual lines that act as "guidelines" for the possible solution curves. Pick any point not on one of the lines. If the point is near the line with basis $v_1 = \begin{bmatrix} 1 \\ 5 \end{bmatrix}$, curves will flow away from the origin (the associated eigenvalue is $\lambda_1 = 5$). If the point is near the line with basis $v_2 = \begin{bmatrix} 1 \\ -3 \end{bmatrix}$ (associated eigenvalue $\lambda_2 = -3$), the curve will flow in toward the origin, but then swing away and flow away from the origin. If the chosen point is on the line with basis $v_2 = \begin{bmatrix} 1 \\ -3 \end{bmatrix}$, it will flow into the origin. All others will flow away. The origin is an unstable node.

In this case, all curves will trend away from the origin (which is a trivial solution to the system). This would be an unstable node.

The eigenvalues may be real but not integers:

Example 26.2: Solve $\mathbf{x}' = \begin{bmatrix} 2 & 1 \\ 3 & 2 \end{bmatrix} \mathbf{x}$. (This matrix appeared in Example 25.3)

Solution: The eigenvalues are $\lambda_1 = 2 + \sqrt{3}$ and $\lambda_2 = 2 - \sqrt{3}$, and the eigenvectors are $v_1 = \begin{bmatrix} 1 \\ \sqrt{3} \end{bmatrix}$ and $v_2 = \begin{bmatrix} 1 \\ -\sqrt{3} \end{bmatrix}$. Thus, the general solution is

$$\mathbf{x}(t) = C_1 \begin{bmatrix} 1 \\ \sqrt{3} \end{bmatrix} e^{(2+\sqrt{3})t} + C_2 \begin{bmatrix} 1 \\ -\sqrt{3} \end{bmatrix} e^{(2-\sqrt{3})t}.$$

Larger systems are solved the same way.

Example 26.3: Solve $\mathbf{x}' = \begin{bmatrix} 2 & 1 & -1 \\ 0 & 1 & 2 \\ 0 & 0 & 3 \end{bmatrix} \mathbf{x}$. (The matrix from Example 25.5)

Solution: The eigenvalues are $\lambda_1 = 2$, $\lambda_2 = 1$ and $\lambda_3 = 3$, and their corresponding eigenvectors are $v_1 = \begin{bmatrix} 1 \\ 0 \\ 0 \end{bmatrix}$, $v_2 = \begin{bmatrix} 1 \\ -1 \\ 0 \end{bmatrix}$ and $v_3 = \begin{bmatrix} 0 \\ 1 \\ 1 \end{bmatrix}$. Thus, the general solution is

$$\mathbf{x}(t) = C_1 \begin{bmatrix} 1 \\ 0 \\ 0 \end{bmatrix} e^{2t} + C_2 \begin{bmatrix} 1 \\ -1 \\ 0 \end{bmatrix} e^{t} + C_3 \begin{bmatrix} 0 \\ 1 \\ 1 \end{bmatrix} e^{3t}.$$

Example 26.4: Solve the IVP $\mathbf{x}' = \begin{bmatrix} 4 & 3 \\ 2 & 3 \end{bmatrix} \mathbf{x}$, where $\mathbf{x}(0) = \begin{bmatrix} 1 \\ 2 \end{bmatrix}$.

Solution: The eigenvalues of $\begin{bmatrix} 4 & 3 \\ 2 & 3 \end{bmatrix}$ are $\lambda_1 = 6$ and $\lambda_2 = 1$. The eigenvectors are $v_1 = \begin{bmatrix} 3 \\ 2 \end{bmatrix}$ and $v_2 = \begin{bmatrix} 1 \\ -1 \end{bmatrix}$. Thus, the general solution is

$$\mathbf{x}(t) = C_1 \begin{bmatrix} 3 \\ 2 \end{bmatrix} e^{6t} + C_2 \begin{bmatrix} 1 \\ -1 \end{bmatrix} e^{t}.$$

To find the constants, let $t = 0$. This gives $\begin{bmatrix} 1 \\ 2 \end{bmatrix} = C_1 \begin{bmatrix} 3 \\ 2 \end{bmatrix} + C_2 \begin{bmatrix} 1 \\ -1 \end{bmatrix}$. This is a system $\begin{matrix} 1 = 3C_1 + C_2 \\ 2 = 2C_1 - C_2 \end{matrix}$. Solving it, we find that $C_1 = \frac{3}{5}$ and $C_2 = -\frac{4}{5}$. Thus, the solution is

$$\mathbf{x}(t) = (3/5) \begin{bmatrix} 3 \\ 2 \end{bmatrix} e^{6t} - (4/5) \begin{bmatrix} 1 \\ -1 \end{bmatrix} e^{t}.$$

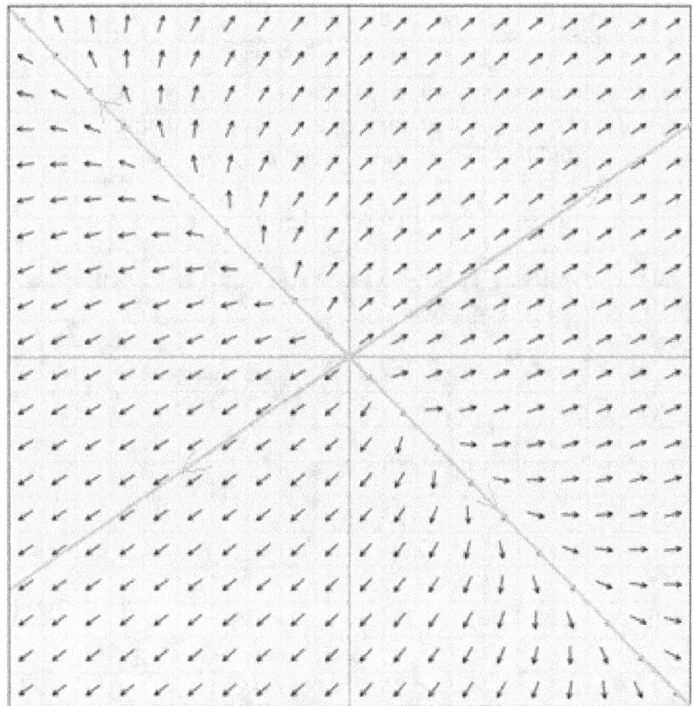

Scale: each gridline is 0.5 unit. Courtesy https://aeb019.hosted.uark.edu/pplane.html.

Above is the phase portrait of

$$\mathbf{x}' = \begin{bmatrix} 2 & 1 \\ 3 & 2 \end{bmatrix} \mathbf{x} \qquad \text{(Example 25.3)}.$$

Note the eigenvectors $\begin{bmatrix} 1 \\ -1 \end{bmatrix}$ and $\begin{bmatrix} 3 \\ 2 \end{bmatrix}$ each form a basis for individual lines (eigenspaces) that act as "guidelines" for the possible solution curves. Because both eigenvalues are positive, all curves will trend away from the origin (which is a trivial solution to the system). This is an unstable node.

Section 27
Systems: Complex Eigenvalues

Recall that $e^{ait} = \cos at + i \sin at$. We will use this identity when solving systems of differential equations with constant coefficients in which the eigenvalues are complex.

We start with a simple case discussed much earlier.

Example 27.1: Solve $y'' + y = 0$ by using auxiliary polynomials, and then again using matrices.

Solution: The auxiliary polynomial is $r^2 + 1 = 0$. This gives $r = \pm i$ so that $e^{it} = \cos t + i \sin t$ is a solution, implying that $\cos t$ and $\sin t$ are the two real-valued solutions, and that the general solution is

$$y = C_1 \cos t + C_2 \sin t.$$

Using matrices, we let $x_1(t) = y$ so that $x_2(t) = x_1'(t) = y'$ and that $x_2'(t) = y''$. So now we have two equations:

$$x_1'(t) = x_2(t)$$
$$x_2'(t) + x_1(t) = 0$$

Isolate the two derivatives to the left side:

$$x_1'(t) = x_2(t)$$
$$x_2'(t) = -x_1(t)$$

It may help to "zero fill" to see the structure better:

$$x_1'(t) = 0x_1(t) + 1x_2(t)$$
$$x_2'(t) = -1x_1(t) + 0x_2(t)$$

In matrix form, the equivalent differential equation is

$$\begin{bmatrix} x_1'(t) \\ x_2'(t) \end{bmatrix} = \begin{bmatrix} 0 & 1 \\ -1 & 0 \end{bmatrix} \begin{bmatrix} x_1(t) \\ x_2(t) \end{bmatrix}.$$

We find the eigenvectors: $\det\begin{bmatrix} 0-\lambda & 1 \\ -1 & 0-\lambda \end{bmatrix} = 0$ gives $\lambda^2 + 1 = 0$, with solutions $\lambda = \pm i$. We study $\lambda_1 = i$ first.

The eigenvector for $\lambda_1 = i$ is $\begin{bmatrix} a \\ b \end{bmatrix}$ such that

$$\begin{bmatrix} 0-i & 1 \\ -1 & 0-i \end{bmatrix}\begin{bmatrix} a \\ b \end{bmatrix} = \begin{bmatrix} 0 \\ 0 \end{bmatrix}.$$

This simplifies to $\begin{bmatrix} -i & 1 \\ -1 & -i \end{bmatrix}\begin{bmatrix} a \\ b \end{bmatrix} = \begin{bmatrix} 0 \\ 0 \end{bmatrix}$. Multiplying the top row with $\begin{bmatrix} a \\ b \end{bmatrix}$ gives $-ia + b = 0$. If we let $a = 1$, then $b = i$. Therefore, the eigenvector for $\lambda_1 = i$ is $v_1 = \begin{bmatrix} 1 \\ i \end{bmatrix}$. The eigenvector for $\lambda_2 = -i$ is found in a similar manner and is $v_2 = \begin{bmatrix} 1 \\ -i \end{bmatrix}$. Thus, the solution in Complex form is

$$\mathbf{x}(t) = C_1 \begin{bmatrix} 1 \\ i \end{bmatrix} e^{it} + C_2 \begin{bmatrix} 1 \\ -i \end{bmatrix} e^{-it}.$$

At this moment, the constants C_1 and C_2 can be ignored. They'll reappear momentarily.

The first term is $\begin{bmatrix} 1 \\ i \end{bmatrix} e^{it}$. Since $e^{it} = \cos t + i \sin t$, we have

$$\begin{bmatrix} 1 \\ i \end{bmatrix} e^{it} = \begin{bmatrix} 1 \\ i \end{bmatrix}(\cos t + i \sin t)$$

$$= \begin{bmatrix} \cos t + i \sin t \\ i(\cos t + i \sin t) \end{bmatrix}$$

$$= \begin{bmatrix} \cos t + i \sin t \\ -\sin t + i \cos t \end{bmatrix}$$

$$= \begin{bmatrix} \cos t \\ -\sin t \end{bmatrix} + i \begin{bmatrix} \sin t \\ \cos t \end{bmatrix}.$$

In the last step, we split the columns of the previous matrix into two columns and factored the i to the front. What remains are two linearly independent column vectors, both potentially solutions to the differential equation. We now perform the same analysis for the second term, $\begin{bmatrix} 1 \\ -i \end{bmatrix} e^{-it}$:

$$\begin{bmatrix} 1 \\ -i \end{bmatrix} e^{-it} = \begin{bmatrix} 1 \\ -i \end{bmatrix}(\cos(-t) + i \sin(-t))$$

$$= \begin{bmatrix} 1 \\ -i \end{bmatrix}(\cos t - i \sin t)$$

$$= \begin{bmatrix} \cos t - i\sin t \\ -i(\cos t - i\sin t) \end{bmatrix}$$

$$= \begin{bmatrix} \cos t - i\sin t \\ -\sin t - i\cos t \end{bmatrix}$$

$$= \begin{bmatrix} \cos t \\ -\sin t \end{bmatrix} + i \begin{bmatrix} -\sin t \\ -\cos t \end{bmatrix}.$$

The second column is just a scalar multiple of $\begin{bmatrix} \sin t \\ \cos t \end{bmatrix}$ from the first solution form. Thus, the solution is $\mathbf{x}(t) = C_1 \begin{bmatrix} \cos t \\ -\sin t \end{bmatrix} + C_2 \begin{bmatrix} \sin t \\ \cos t \end{bmatrix}.$

Note that using matrices produced the same solution as did using auxiliary polynomials, but matrices can be extended to handle more complicated forms, especially those involving much higher orders of the derivative.

Example 27.2: Solve $\mathbf{x}' = \begin{bmatrix} 3 & -2 \\ 4 & -1 \end{bmatrix} \mathbf{x}.$

Solution: Find the eigenvalues first: $\det \begin{bmatrix} 3-\lambda & -2 \\ 4 & -1-\lambda \end{bmatrix} = 0$ gives

$$\lambda^2 - 2\lambda + 5 = 0$$

After simplification. Using the quadratic formula to solve, it provides two eigenvalues, $\lambda_1 = 1 + 2i$ and $\lambda_2 = 1 - 2i$.

The eigenvector for $\lambda_1 = 1 + 2i$ is $\begin{bmatrix} a \\ b \end{bmatrix}$ such that

$$\begin{bmatrix} 3 - (1+2i) & -2 \\ 4 & -1-(1+2i) \end{bmatrix} \begin{bmatrix} a \\ b \end{bmatrix} = \begin{bmatrix} 0 \\ 0 \end{bmatrix}.$$

This simplifies to $\begin{bmatrix} 2-2i & -2 \\ 4 & -2-2i \end{bmatrix} \begin{bmatrix} a \\ b \end{bmatrix} = \begin{bmatrix} 0 \\ 0 \end{bmatrix}$. Multiplying the top row with $\begin{bmatrix} a \\ b \end{bmatrix}$ gives $(2-2i)a - 2b = 0$. If we let $a = 1$, then $b = 1 - i$. Thus, the eigenvector for $\lambda_1 = 1 + 2i$ is $v_1 = \begin{bmatrix} 1 \\ 1-i \end{bmatrix}$. The eigenvector for $\lambda_2 = 1 - 2i$ is found in a similar way and is $v_2 = \begin{bmatrix} 1 \\ 1+i \end{bmatrix}$.

Don't forget that since the above matrix is singular, the two rows are multiples of one another, so that it is sufficient to multiply one row to $\begin{bmatrix} a \\ b \end{bmatrix}$. Multiplying the bottom row to $\begin{bmatrix} a \\ b \end{bmatrix}$ would provide the same eigenvector (possibly a scalar multiple, which is sufficient).

The solution of the differential equation in Complex form is

$$x(t) = C_1 \begin{bmatrix} 1 \\ 1-i \end{bmatrix} e^{(1+2i)t} + C_2 \begin{bmatrix} 1 \\ 1+i \end{bmatrix} e^{(1-2i)t}.$$

We now rewrite $x(t) = C_1 \begin{bmatrix} 1 \\ 1-i \end{bmatrix} e^{(1+2i)t} + C_2 \begin{bmatrix} 1 \\ 1+i \end{bmatrix} e^{(1-2i)t}$ in real form using the identity $e^{nit} = \cos nt + i \sin nt$. The leading constants can be ignored for the following analysis.

Consider the first term: $\begin{bmatrix} 1 \\ 1-i \end{bmatrix} e^{(1+2i)t}$.

Recall that $e^{(1+2i)t} = e^t e^{2it} = e^t(\cos 2t + i \sin 2t)$. thus, we have

$$\begin{bmatrix} 1 \\ 1-i \end{bmatrix} e^{(1+2i)t} = \begin{bmatrix} 1 \\ 1-i \end{bmatrix} e^t(\cos 2t + i \sin 2t)$$

$$= e^t \begin{bmatrix} \cos 2t + i \sin 2t \\ (1-i)(\cos 2t + i \sin 2t) \end{bmatrix}.$$

Expand the second row by multiplication, and simplify:

$$e^t \begin{bmatrix} \cos 2t + i \sin 2t \\ \cos 2t + i \sin 2t - i \cos 2t - i^2 \sin 2t \end{bmatrix}$$

$$= e^t \begin{bmatrix} \cos 2t + i \sin 2t \\ \cos 2t + \sin 2t + i(\sin 2t - \cos 2t) \end{bmatrix}.$$

Doing the same with the second term of the solution, $\begin{bmatrix} 1 \\ 1+i \end{bmatrix} e^{(1-2i)t}$, gives a scalar multiple of the results above from the first term. Once terms are combined and constants renamed, we end with the same result. Thus, it is sufficient to perform this process just once, as we have done above.

Now, "stack" the terms into two columns, one real and one imaginary:

$$e^t \begin{bmatrix} \cos 2t & +i \sin 2t \\ \cos 2t + \sin 2t & +i(\sin 2t - \cos 2t) \end{bmatrix}.$$

Recall that if $u(t) + iv(t)$ are solutions of a homogeneous ODE, then so are $u(t)$ and $v(t)$. We can drop the imaginary unit now. The solution of the differential equation $\mathbf{x}' = \begin{bmatrix} 3 & -2 \\ 4 & -1 \end{bmatrix} \mathbf{x}$ is:

$$\mathbf{x}(t) = C_1 e^t \begin{bmatrix} \cos 2t \\ \cos 2t + \sin 2t \end{bmatrix} + C_2 e^t \begin{bmatrix} \sin 2t \\ \sin 2t - \cos 2t \end{bmatrix}.$$

Now it is appropriate to attach the two leading generic constants.

We verify that the two solutions above are linearly independent by checking its Wronskian:

$$W = \det \begin{bmatrix} e^t \cos 2t & e^t \sin 2t \\ e^t(\cos 2t + \sin 2t) & e^t(\sin 2t - \cos 2t) \end{bmatrix}$$

$$= e^{2t} \cos 2t \,(\sin 2t - \cos 2t) - e^{2t} \sin 2t \,(\cos 2t + \sin 2t)$$

$$= e^{2t} \cos 2t \sin 2t - e^{2t} \cos^2 2t - e^{2t} \sin 2t \cos 2t - e^{2t} \sin^2 2t$$

$$= -e^{2t} \cos^2 2t - e^{2t} \sin^2 2t$$

$$= -e^{2t} \underbrace{(\cos^2 2t + \sin^2 2t)}_{1}$$

$$= -e^{2t}.$$

Since the Wronskian $-e^{2t}$ is never zero, these are linearly independent solutions.

Phase Portraits

The phase portraits of a system of differential equations with complex eigenvalues fall into three categories, depending on the sign of the value of the real part of the complex eigenvalue. A solution curve will "follow" the arrows, so to speak. Assume the eigenvalues have the form $\lambda = a \pm bi$.

If $a > 0$, then the direction curves trend away from the origin. The origin is called an unstable spiral point.

If $a < 0$, then the direction curves trend toward the origin. The origin is called a stable spiral point.

If $a = 0$, then the direction curves form concentric ellipses around the origin.

Example 27.3: Sketch the phase portrait of the solutions of $\mathbf{x}' = \begin{bmatrix} 3 & -2 \\ 4 & -1 \end{bmatrix} \mathbf{x}$. (See Example 27.2)

Solution: The eigenvalues are $\lambda = 1 \pm 2i$. Since $a > 0$, the solution curves spiral away from the origin. The origin is an unstable spiral point. Its phase portrait is

Scale: each gridline is 0.5 unit. Courtesy https://aeb019.hosted.uark.edu/pplane.html.

Example 27.4: Sketch the phase portrait of the solutions of $\mathbf{x}' = \begin{bmatrix} -1 & -1 \\ 2 & -1 \end{bmatrix} \mathbf{x}$.

Solution: The eigenvalues are $\lambda = -1 \pm \sqrt{2}i$. Since $a < 0$, the solution curves spiral toward the origin. The origin is a stable spiral point. Its phase portrait is

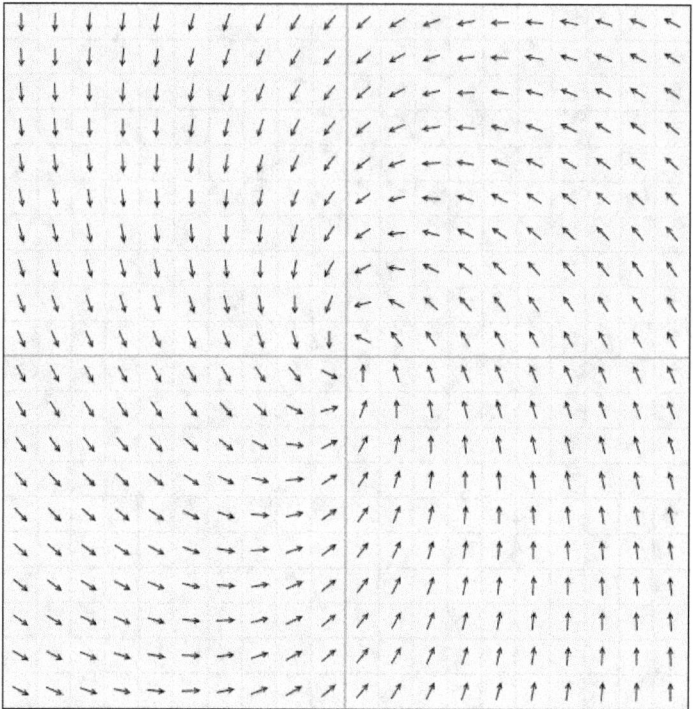

Scale: each gridline is 0.5 unit. Courtesy https://aeb019.hosted.uark.edu/pplane.html.

The solution curves approach the origin as a limit, in ever smaller revolutions (orbits).

Example 27.5: Sketch the phase portrait of the solutions of $\mathbf{x}' = \begin{bmatrix} 0 & -2 \\ 4 & 0 \end{bmatrix} \mathbf{x}$.

Solution: The eigenvalues are $\lambda = \pm 2\sqrt{2}i$. Since $a = 0$, the solution curves follow elliptical paths around the origin. Its phase portrait is

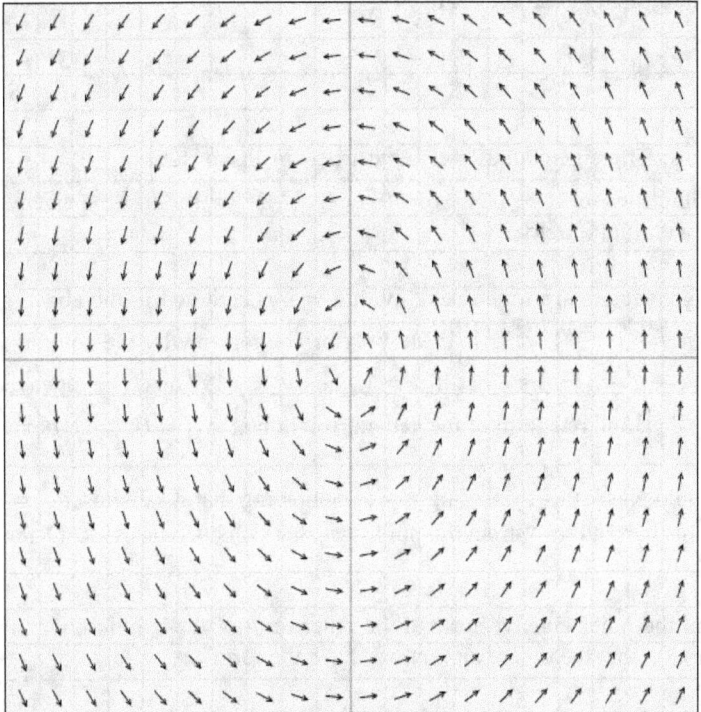

Scale: each gridline is 0.5 unit. Courtesy https://aeb019.hosted.uark.edu/pplane.html.

Section 28
Systems: Real Eigenvalues of Repeated Multiplicity

We look at a case where the eigenvalue(s) of a system of differential equations have multiplicity greater than 1.

Example 28.1: Solve $\mathbf{x}' = \begin{bmatrix} 3 & -1 \\ 1 & 5 \end{bmatrix} \mathbf{x}$. For convenience, let $A = \begin{bmatrix} 3 & -1 \\ 1 & 5 \end{bmatrix}$.

Solution: The eigenvalues are found first. We have $\det \begin{bmatrix} 3-\lambda & -1 \\ 1 & 5-\lambda \end{bmatrix} = 0$. This simplifies to $\lambda^2 - 8\lambda + 16 = 0$, or $(\lambda - 4)^2 = 0$. Thus, the eigenvalue is $\lambda = 4$ with multiplicity 2.

The eigenvectors are found next. With $\lambda = 4$ substituted in the above matrix, we have $\begin{bmatrix} -1 & -1 \\ 1 & 1 \end{bmatrix} \begin{bmatrix} a \\ b \end{bmatrix} = \begin{bmatrix} 0 \\ 0 \end{bmatrix}$. Multiplying the top row by the column vector, we have $-a - b = 0$ so that if $a = 1$, then $b = -1$. Thus, one eigenvector is $v_1 = \begin{bmatrix} 1 \\ -1 \end{bmatrix}$, and one term of the general solution is $x_{(1)} = C_1 \begin{bmatrix} 1 \\ -1 \end{bmatrix} e^{4t}$.

We would check this by finding $x'_{(1)}$ and showing that it satisfies $x'_{(1)} = Ax_{(1)}$. Keep this in mind for "structural purposes" that will become evident a few steps ahead.

To find the other eigenvector and the other term of the solution, we make an assumption that the other solution will have the form

$$x_{(2)} = (mt + n)e^{4t} = mte^{4t} + ne^{4t},$$

where m and n are two 2×1 vectors to be determined. Differentiating, we have

$$x'_{(2)} = 4mte^{4t} + me^{4t} + 4ne^{4t}.$$

The assumption is that $x'_{(2)} = Ax_{(2)}$, similar to above with $x_{(1)}$. Thus, substituting the expressions for $x'_{(2)}$ and $x_{(2)}$, we have

$$4mte^{4t} + me^{4t} + 4ne^{4t} = A(mte^{4t} + ne^{4t}).$$

Equating by like terms, we have

$$4mte^{4t} = Amte^{4t} \quad \text{and} \quad me^{4t} + 4ne^{4t} = Ane^{4t}.$$

The first equation simplifies to $4m = Am$. Since we know 4 is an eigenvalue of A, then the first statement is true only if vector m is the eigenvector we found earlier, $v_1 = \begin{bmatrix} 1 \\ -1 \end{bmatrix}$. Thus, we have $m = \begin{bmatrix} 1 \\ -1 \end{bmatrix}$.

In the second equation, rearrange terms to place the n terms on one side, and write v_1 in place of m:

$$(An - 4n)e^{4t} = v_1 e^{4t}, \text{ which implies that } An - 4In = v_1.$$

Thus, we are solving the equation

$$\begin{bmatrix} -1 & -1 \\ 1 & 1 \end{bmatrix} n = \underbrace{\begin{bmatrix} 1 \\ -1 \end{bmatrix}}_{v_1}.$$

Let $n = \begin{bmatrix} n_1 \\ n_2 \end{bmatrix}$. The top row multiplied to n gives $-n_1 - n_2 = 1$. We solve generically for all possible n. If we let $n_1 = k$, then $n_2 = -1 - k$. Thus, vector n is now written

$$n = \begin{bmatrix} k \\ -1 - k \end{bmatrix} = \begin{bmatrix} 0 + 1k \\ -1 - 1k \end{bmatrix} = \begin{bmatrix} 0 \\ -1 \end{bmatrix} + k \begin{bmatrix} 1 \\ -1 \end{bmatrix}.$$

Note that the vector attached to k is v_1. This is a good check of your work. Also, a new vector, $\begin{bmatrix} 0 \\ -1 \end{bmatrix}$, appeared. Call this n_a for the moment.

We now have all the components for the second solution $x_{(2)}$:

$$x_{(2)} = mte^{4t} + ne^{4t} = \underbrace{\begin{bmatrix} 1 \\ -1 \end{bmatrix}}_{m} te^{4t} + \underbrace{\left(\begin{bmatrix} 0 \\ -1 \end{bmatrix} + k \begin{bmatrix} 1 \\ -1 \end{bmatrix} \right)}_{n} e^{4t}.$$

The general solution is

$$x(t) = x_{(1)}(t) + x_{(2)}(t)$$

$$= C_1 \begin{bmatrix} 1 \\ -1 \end{bmatrix} e^{4t} + C_2 \left(\begin{bmatrix} 1 \\ -1 \end{bmatrix} te^{4t} + \left(\begin{bmatrix} 0 \\ -1 \end{bmatrix} + k \begin{bmatrix} 1 \\ -1 \end{bmatrix} \right) e^{4t} \right).$$

When C_2 is multiplied through the parentheses, the term $C_2 k \begin{bmatrix} 1 \\ -1 \end{bmatrix} e^{4t}$ is linearly dependent with the first solution, $C_1 \begin{bmatrix} 1 \\ -1 \end{bmatrix} e^{4t}$. Thus, the two can be combined (in the sense of adding like terms) to create $(C_1 + C_2 k) \begin{bmatrix} 1 \\ -1 \end{bmatrix} e^{4t}$, or just $C_1 \begin{bmatrix} 1 \\ -1 \end{bmatrix} e^{4t}$, where "new" C_1 is $C_1 + C_2 k$.

The general solution, now simplified, is

$$\mathbf{x}(t) = C_1 \begin{bmatrix} 1 \\ -1 \end{bmatrix} e^{4t} + C_2 \left(\begin{bmatrix} 1 \\ -1 \end{bmatrix} t e^{4t} + \begin{bmatrix} 0 \\ -1 \end{bmatrix} e^{4t} \right).$$

To summarize, if $\mathbf{x}' = A\mathbf{x}$ is a system where A is a 2×2 matrix with an eigenvalue λ of multiplicity 2 and associated eigenvector v_1, then the general solution is

$$x(t) = C_1 v_1 e^{\lambda t} + C_2 [v_1 t e^{\lambda t} + n_a e^{\lambda t}],$$

where n is found by solving $An - \lambda I n = v_1$ and it has the form $n = n_a + k v_1$.

The phase portrait of this system follows.

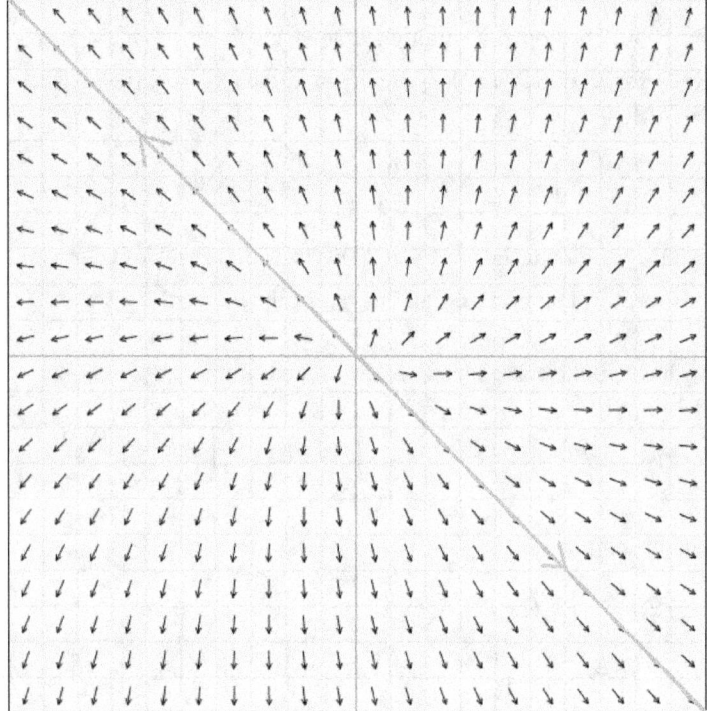

Scale: each gridline is 0.5 unit. Courtesy https://aeb019.hosted.uark.edu/pplane.html.

In this case, the eigenvalue is positive, and its eigenvector, $\begin{bmatrix} 1 \\ -1 \end{bmatrix}$ is the basis of the line shown above. All curves flow away from the origin. In this case, the origin is called an improper unstable node (had the curves flowed inward, it would be an improper stable node).

Addendum

Free Extra Stuff!

The following topics are sometimes covered in an introductory differential equations course. An abbreviated discussion for each is presented here.

The Imaginary Unit i, the Complex Field and Roots of Unity

Partial Fraction Decomposition

Variation of Parameters

Exact Equations

Series Solutions

The Imaginary Unit *i*, the Complex Field and Roots of Unity

The Real numbers are composed of the positive and negative integers, the rational numbers, and the irrational numbers. Rational numbers are expressible by the ratio of two integers, and include the integers as a subset, since, for example, the integer 4 is expressible as the ratio 4/1. The decimal expansion of a rational number always terminates or repeats the same bloc of values forever. For example, 4 = 4.0, 9/8 = 1.125, and 3/11 = 0.272727… . Irrational numbers are not expressible as the ratio of two integers, and their decimal expansions never terminate nor repeat the same bloc ad infinitum. Common examples are π, $\sqrt{2}$ and e. Roots of even index always require the radicand be non-negative. The arithmetic operations of addition, subtraction, multiplication and division (not by 0) are all well-defined over the Real numbers. The structure is internally consistent and self-contained. The Real numbers are an example of a mathematical **field**. The Real numbers are often expressed as a single line on which all Real numbers have a unique position on that line. There are no holes, and no position is occupied by different Real numbers (after simplification).

However, an expression like $\sqrt{-1}$ is not defined in the Real number field. There is no real number that when squared, gives -1. To resolve this conundrum, a new number is defined, the imaginary unit $i = \sqrt{-1}$. This number is clearly not a Real number. When composed with Real numbers in the form $a + bi$, where a and b are Real numbers, the result is called a Complex number. The arithmetic operations are well-defined, and the Complex numbers also form a field. It is usually expressed by a coordinate axis system, where the horizontal axis represents the Real numbers, and the vertical axis represents the Imaginary numbers, and a Complex number such as $2 + 3i$ is represented by the point (2,3). The Complex numbers form an extension of the Real numbers in that the Real numbers are a subset of the Complex numbers. In a sense, they are "higher dimensional" numbers.

When raised to an integer power n, the values of i^n form a pattern:

$$i^1 = i, \quad i^2 = -1, \quad i^3 = -i, \quad i^4 = 1.$$

The pattern repeats:

$$i^5 = i, \quad i^6 = -1, \quad i^7 = -i, \quad i^8 = 1, \quad \text{and so on.}$$

Negative integer powers also follow this pattern:

$$i^{-3} = i, \quad i^{-2} = -1, \quad i^{-1} = -i, \quad i^0 = 1.$$

A remarkable identity known as Euler's Formula relates the natural base e, the imaginary unit i, and the principal trigonometric functions cosine and sine:

$$e^{it} = \cos t + i \sin t.$$

This is found by expanding each function by its Maclaurin series and making substitutions (this is discussed in Section 12).

The n Roots of Unity & deMoivre's Formula

A root of unity is any number (Real or Complex) such that when raised to a positive integer power, gives 1.

In the Real numbers alone, the roots of unity are 1, when the root index is even, and -1 and 1 when the root index is odd. However, in the Complex field, more roots of unity can exist. For example, there are four fourth roots of unity: $1, -1, i$ and $-i$. This means that $1^4 = 1, (-1)^4 = 1, i^4 = 1$ and $(-i)^4 = 1$. In general, in the Complex field, there are n nth roots of unity for any positive integer n. That means there are 3 third roots of 1, 5 fifth-roots of 1, and so on.

Since $e^{it} = \cos t + i \sin t$, raising both sides by a positive integer n, we have

$$(e^{it})^n = (\cos t + i \sin t)^n, \text{ which gives } e^{int} = \cos nt + i \sin nt.$$

Simplified, we have

$$(\cos t + i \sin t)^n = \cos nt + i \sin nt.$$

This is one common form of writing deMoivre's Formula. Although this formula requires that n be a positive integer, t can be any real number. Thus, an expression such as $e^i = (e^{0.5i})^2 = (\cos 0.5 + i \sin 0.5)^2$ "makes sense", which can be checked by a calculator. Among other things, it allows a way to define imaginary exponents.

Extending this theme, we can now restate deMoivre's Formula as

$$e^{2\pi ki/n} = \cos\left(\frac{2\pi k}{n}\right) + i \sin\left(\frac{2\pi k}{n}\right), \quad \text{for } k = 0, 1, 2, \ldots, n-1.$$

Since the period of the cosine function and sine function is 2π, the n-value subdivides the unit circle into n equal subdivisions, starting at 1, and proceeding counterclockwise, the corresponding point on the circle representing a root of unity (where the "y" value is the imaginary part of the Complex root).

For example, when $n = 2$, this subdivides the unit circle into 2 equal parts. When $k = 0$, this is the angle 0 and corresponds to the point (1,0) on the unit circle, corresponding to the Complex number $1 + 0i$, or equivalently, the Real number

1. When $k = 1$, this is the angle π radians and corresponds to the point $(-1,0)$ on the unit circle, corresponding to the Complex number $-1 + 0i$, or equivalently, the Real number -1. These are the 2 second roots of unity and should not come as a surprise.

It is also easy to visualize that the 4 fourth roots of unity are 1, i, -1 and $-i$. These correspond to the four evenly spaced points around the unit circle, each separated by $90°$, or $\pi/2$ radians.

Let's explore the case when $n = 3$, so that $k = 0$, 1, and 2. We have

$$k = 0: \quad \cos 0 + i \sin 0 = 1 + 0i = 1;$$
$$k = 1: \quad \cos\frac{2\pi}{3} + i \sin\frac{2\pi}{3} = -\frac{1}{2} + \frac{\sqrt{3}}{2}i;$$
$$k = 2: \quad \cos\frac{4\pi}{3} + i \sin\frac{4\pi}{3} = -\frac{1}{2} - \frac{\sqrt{3}}{2}i.$$

These are the three cube (or third) roots of 1:

$$1^3 = 1, \quad \left(-\frac{1}{2} + \frac{\sqrt{3}}{2}i\right)^3 = 1, \quad \left(-\frac{1}{2} - \frac{\sqrt{3}}{2}i\right)^3 = 1.$$

These three numbers are expressed as three equally spaced points on the unit circle, starting at $(1,0)$ and proceeding counterclockwise.

In general, the n nth-roots of unity are expressible as n evenly-spaced points on the unit circle, always starting at $(1,0)$ (the value 1 is the primitive, or primary, nth-root of 1). Visually, these points are the vertices of a regular n-gon inscribed within the circle, one vertex at $(1,0)$.

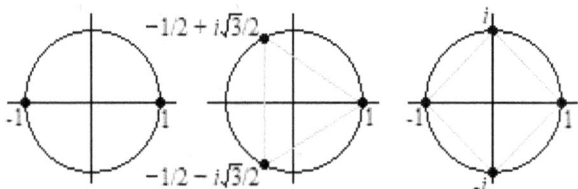

A visual representation of the 2 square roots, the 3 cube roots and the 4 fourth roots of unity.

As hinted earlier, we can now define imaginary powers in a meaningful manner using the Euler and deMoivre formulas. Let a be a positive real number, and we consider the meaning of a^i. We use the identity $a = e^{\ln a}$, so that $a^i = e^{i \ln a}$. For example, $3^i = e^{i \ln 3} = \cos(\ln 3) + i \sin(\ln 3)$. This can be verified on a calculator.

Lastly, what is i^i? Using the Euler Formula $e^{it} = \cos t + i \sin t$, let $t = \frac{\pi}{2}$ so that we have $e^{i(\pi/2)} = \cos \pi/2 + i \sin \pi/2$, or $e^{i(\pi/2)} = 0 + 1i = i$. Now, raise both sides by i:

$$i^i = \left(e^{i(\pi/2)}\right)^i = e^{i^2(\pi/2)} = e^{-\pi/2}.$$

Interestingly, i^i is a Real number, $e^{-\pi/2}$. A calculator confirms this. What does this mean? You can ponder on that.

Partial Fraction Decomposition

When summing two or more expressions written in fraction (or rational) form, we must first rewrite each using a common denominator. For example, consider the sum $\frac{2}{x+3} + \frac{1}{x-5}$. The arithmetic is as follows:

$$\frac{2}{x+3} + \frac{1}{x-5} = \frac{\overbrace{2(x-5) + 1(x+3)}^{\text{adjust numerators}}}{\underbrace{(x+3)(x-5)}_{\text{common denominator}}} = \frac{2x - 10 + x + 3}{(x+3)(x-5)} = \frac{3x - 7}{(x+3)(x-5)}$$

The process in reverse is called **partial fraction decomposition**. For example, if given $\frac{3x-7}{(x+3)(x-5)}$, then there is a systematic way to "decompose" this as the sum of two "smaller" rational expressions. We should be able to show that $\frac{3x-7}{(x+3)(x-5)} = \frac{2}{x+3} + \frac{1}{x-5}$ after the procedure is completed. By "smaller", we mean that the degree of each of the denominators is smaller than that of the original expression.

We will study polynomials that factor into linear factors (of the form $ax + b$), or irreducible quadratics of the form $ax^2 + bx + c$ that do not factor over the Reals.

Case I: The denominator factors into two or more distinct linear factors, each of multiplicity 1 (that is, each factor is not raised to a power of 2 or higher).

Example: Decompose $\frac{3x-7}{(x+3)(x-5)}$ as the sum of two or more summands.

Solution: Since the denominator is already factored, and each factor is linear (degree 1), we assign temporary numerators to each using capital letters A, B, C, etc.:

$$\frac{3x-7}{(x+3)(x-5)} = \frac{A}{x+3} + \frac{B}{x-5}$$

Now, recompose the right side by finding the common denominator and adjusting the numerators accordingly:

$$\frac{A}{x+3} + \frac{B}{x-5} = \frac{A(x-5) + B(x+3)}{(x+3)(x-5)}.$$

We now compare the numerators of the original expression, and the new one with the temporary constants:

$$3x - 7 = A(x-5) + B(x+3).$$

There are two ways to determine the values of A and B. The formal method is to clear parentheses on the right side and develop a system by comparing coefficients:

$$\begin{aligned} 3x - 7 &= A(x-5) + B(x+3) \\ &= Ax - 5A + Bx + 3B \\ &= (A+B)x + (-5A + 3B). \end{aligned}$$

The coefficient of x on the left side of the equation is 3, and on the right side is $A + B$. Thus, we have $A + B = 3$. Similarly, the constant on the left side is -7 and on the right side is $-5A + 3B$. Thus, we have $-5A + 3B = -7$. This is a system of two equations of two unknowns:

$$\begin{aligned} A + B &= 3 \\ -5A + 3B &= -7. \end{aligned}$$

This system can be solved in many ways, through substitution, or using matrix row operations. Any method, if properly performed, works. In this example, we get $A = 2$ and $B = 1$.

Thus, we have shown that

$$\frac{3x-7}{(x+3)(x-5)} = \frac{2}{x+3} + \frac{1}{x-5}.$$

A more heuristic method to determine A and B is to compare the numerators:

$$3x - 7 = A(x-5) + B(x+3).$$

By selecting certain "strategic" values for x, we can isolate A or B and solve for each quickly. For example, when $x = 5$, we have

$$3(5) - 7 = A(5 - 5) + B(5 + 3)$$
$$8 = 8B$$
$$B = 1.$$

Note that the A term vanishes. Similarly, when $x = -3$, we have

$$3(-3) - 7 = A(-3 - 5) + B(-3 + 3)$$
$$-16 = -8A$$
$$A = 2.$$

Here, the B term vanishes, and we found A and B as desired. We again arrive at the same decomposition,

$$\frac{3x - 7}{(x + 3)(x - 5)} = \frac{2}{x + 3} + \frac{1}{x - 5}.$$

Example: Find the partial fraction decomposition for

$$\frac{x^2 + 3x + 4}{x^3 - 4x^2 - 15x + 18}.$$

Solution: The denominator factors:

$$x^3 - 4x^2 - 15x + 18 = (x - 1)(x + 3)(x - 6).$$

Each factor is linear and of multiplicity 1. Thus, we initially decompose the given rational expression into three summands:

$$\frac{x^2 + 3x + 4}{x^3 - 4x^2 - 15x + 18} = \frac{A}{x - 1} + \frac{B}{x + 3} + \frac{C}{x - 6}.$$

The right side is recomposed by finding a common denominator and the numerators adjusted accordingly:

$$\frac{A}{x - 1} + \frac{B}{x + 3} + \frac{C}{x - 6}$$
$$= \frac{A(x + 3)(x - 6) + B(x - 1)(x - 6) + C(x - 1)(x + 3)}{(x - 1)(x + 3)(x - 6)}.$$

The numerators are then compared:

$$x^2 + 3x + 4 = A(x+3)(x-6) + B(x-1)(x-6) + C(x-1)(x+3).$$

Here, it may be faster to use a "heuristic" method to determine A, B and C:

When $x = -3$, the A and C terms vanish, leaving the B term, and we have

$$(-3)^2 + 3(-3) + 4 = B(-3-1)(-3-6)$$
$$9 - 9 + 4 = B(-4)(-9)$$
$$4 = 36B$$
$$B = 1/9.$$

When $x = 6$, the A and B terms vanish:

$$(6)^2 + 3(6) + 4 = C(6-1)(6+3)$$
$$36 + 18 + 4 = C(5)(9)$$
$$58 = 45C$$
$$C = 58/45.$$

When $x = 1$, the B and C terms vanish:

$$(1)^2 + 3(1) + 4 = A(1+3)(1-6)$$
$$1 + 3 + 4 = A(4)(-5)$$
$$8 = -20A$$
$$A = -2/5.$$

Thus, we have shown that

$$\frac{x^2 + 3x + 4}{x^3 - 4x^2 - 15x + 18} = \frac{-2/5}{x-1} + \frac{1/9}{x+3} + \frac{58/45}{x-6}$$

$$= -\frac{2}{5(x-1)} + \frac{1}{9(x+3)} + \frac{58}{45(x-6)}.$$

Case II: One (or more) of the factors of the denominator is an irreducible quadratic. An irreducible quadratic does not factor over the Real numbers. Its solutions will be complex numbers. Using the quadratic formula on the generic quadratic $ax^2 + bx + c$, its discriminant, $b^2 - 4ac$, governs whether the quadratic is reducible (factorable) or not. If the discriminant is negative, then the quadratic is irreducible over the reals.

Example: Find the partial fraction decomposition for
$$\frac{2x-5}{x^3+x-10}.$$

Solution: The denominator factors:
$$x^3+x-10 = (x-2)(x^2+2x+5).$$

The second factor is not reducible into linear factors. Because this factor is of degree 2, its numerator will be of degree 1:
$$\frac{2x-5}{x^3+x-10} = \frac{A}{x-2} + \frac{Bx+C}{x^2+2x+5}.$$

Now recompose the right side:
$$\frac{A}{x-2} + \frac{Bx+C}{x^2+2x+5} = \frac{A(x^2+2x+5)+(Bx+C)(x-2)}{(x-2)(x^2+2x+5)}.$$

The "trick" of choosing strategic x-values won't work as well here. Instead, we clear parentheses on the right side, then recollect terms according to the powers of x, and develop a system:

$$A(x^2+2x+5)+(Bx+C)(x-2) = Ax^2+2Ax+5A+Bx^2-2Bx+Cx-2C.$$
$$= (A+B)x^2+(2A-2B+C)x+(5A-2C).$$

Comparing coefficients to the original numerator $2x-5$, we have a system:
$$A+B = 0$$
$$2A-2B+C = 2$$
$$5A-2C = -5.$$

Using any solution method, we get $A = -1/13$, $B = 1/13$ and $C = 30/13$. Thus,
$$\frac{2x-5}{x^3+x-10} = \frac{-1/13}{x-2} + \frac{(1/13)x+30/13}{x^2+2x+5} = -\frac{1}{13}\left(\frac{1}{x-2} + \frac{x+30}{x^2+2x+5}\right).$$

Case III: A factor may have multiplicity 2 or greater. For example, the quadratic $x^2+6x+9 = (x+3)(x+3) = (x+3)^2$. Note that the linear factor $(x+3)$ appears twice (raised to the second power) and thus has multiplicity 2.

Example: Find the partial fraction decomposition for

$$\frac{3x+1}{x^2+6x+9}.$$

Solution: The denominator factors as $(x+3)^2$. The form of the decomposition is:

$$\frac{3x+1}{x^2+6x+9} = \frac{A}{x+3} + \frac{B}{(x+3)^2}.$$

The solution process is the same: recompose the summands on the right side, then compare numerators:

$$\frac{A}{x+3} + \frac{B}{(x+3)^2} = \frac{A(x+3)+B}{(x+3)^2} = \frac{Ax+3A+B}{(x+3)^2}.$$

The numerators are:

$$3x+1 = Ax+3A+B.$$

Thus, $A=3$ and since $3A+B=1$, we have $9+B=1$, so $B=-8$.

The partial fraction decomposition is complete:

$$\frac{3x+1}{x^2+6x+9} = \frac{3}{x+3} - \frac{8}{(x+3)^2}.$$

Example: Find the partial fraction decomposition for

$$\frac{x^3+2x^2+2}{x^4+2x^3-15x^2}.$$

Solution: The denominator factors:

$$\frac{x^3+2x^2+2}{x^4+2x^3-15x^2} = \frac{x^3+2x^2+2}{x^2(x^2+2x-15)} = \frac{x^3+2x^2+2}{x^2(x+5)(x-3)}.$$

The factor x^2 is a linear factor of multiplicity 2 so it results in two fractional summands, while the factors $(x+5)$ and $(x-3)$ are each multiplicity 1, so they result in one summand each:

$$\frac{x^3+2x^2+2}{x^2(x+5)(x-3)} = \frac{A}{x} + \frac{B}{x^2} + \frac{C}{x+5} + \frac{D}{x-3}.$$

Now get a common denominator and recompose:

$$\frac{A}{x} + \frac{B}{x^2} + \frac{C}{x+5} + \frac{D}{x-3} = \frac{Ax(x+5)(x-3) + B(x+5)(x-3) + Cx^2(x-3) + Dx^2(x+5)}{x^2(x+5)(x-3)}$$

Then equate numerators:

$$x^3 + 2x^2 + 2 = Ax(x+5)(x-3) + B(x+5)(x-3) + Cx^2(x-3) + Dx^2(x+5).$$

When $x = 0$, the A, C and D terms vanish, and we have

$$(0)^3 + 2(0)^2 + 2 = B(0+5)(0-3)$$
$$2 = -15B$$
$$B = -2/15.$$

When $x = 3$, the A, B and C terms vanish:

$$(3)^3 + 2(3)^2 + 2 = d(3)^2(3+5)$$
$$47 = D(9)(8)$$
$$D = 47/72.$$

When $x = -5$, the A, B and D terms vanish:

$$(-5)^3 + 2(-5)^2 + 2 = C(-5)^2(-5-3)$$
$$-73 = -200C$$
$$C = 73/200.$$

There is no convenient way to remove (vanish) the B, C and D terms to determine A. Instead, we choose any other value for x, and now knowing B, C and D, we can solve for A. Let's let $x = 1$.

$$(1)^3 + 2(1)^2 + 2$$
$$= A(1)(1+5)(1-3) + B(1+5)(1-3) + C(1)^2(1-3) + D(1)^2(1+5)$$

Simplify and substitute the values for B, C and D:

$$5 = A(1)(6)(-2) - \frac{2}{15}(6)(-2) + \frac{73}{200}(1)(-2) + \frac{47}{72}(1)^2(6).$$

$$\frac{16}{75} = -12A$$

$$A = -\frac{4}{225}.$$

We have now separated the original rational expression into its four summands:

$$\frac{x^3 + 2x^2 + 2}{x^4 + 2x^3 - 15x^2} = -\frac{4}{225x} - \frac{2}{15x^2} + \frac{73}{200(x+5)} + \frac{47}{72(x-3)}.$$

Example: Find the partial fraction decomposition for

$$\frac{x^4 + 2}{x^5 + 2x^4 + 10x^3}.$$

Solution: The denominator factors: $x^5 + 2x^4 + 10x^3 = x^3(x^2 + 2x + 10)$. We have a linear factor s, multiplicity 3, and an irreducible quadratic factor. The partial fraction decomposition has the form

$$\frac{x^4 + 2}{x^3(x^2 + 2x + 10)} = \frac{A}{x} + \frac{B}{x^2} + \frac{C}{x^3} + \frac{Dx + E}{x^2 + 2x + 10}.$$

The right side is recomposed:

$$\frac{A}{x} + \frac{B}{x^2} + \frac{C}{x^3} + \frac{Dx + E}{x^2 + 2x + 10} = \frac{(Ax^2 + Bx + C)(x^2 + 2x + 10) + (Dx + E)x^3}{x^3(x^2 + 2x + 10)}$$

To save space, note that A, B and C will each be multiplied by some power of x, as well as by the quadratic factor.

It does not appear that choosing "strategic" values for x will help determine the values of A through E easily. Instead, we multiply out the entire numerator, but even then, this can be done strategically. For example, there will be an x^4 term. Looking at the numerator, this will only occur when Ax^2 is multiplied by x^2, and when Dx is multiplied by x^3. This means we will get $Ax^4 + Dx^4$, which establishes that $A + D = 1$ since 1 is the coefficient of the x^4 term in the original numerator.

We follow the same line of logic for the other powers of x:

For x^3, these will occur when Ax^2 is multiplied by $2x$, when Bx is multipled by x^2, and when E is multiplied by x^3. We get $2Ax^3 + Bx^3 + Ex^3$, and since there is no x^3 term in the original numerator, we have that $2A + B + E = 0$.

For x^2, the terms will be $10Ax^2 + Cx^2 + 2Bx^2$. As there is no x^2 term in the original numerator, we have that $10A + 2B + C = 0$.

For x, the terms are $10Bx + 2Cx$, which establishes that $10B + 2C = 0$.

For the constants, there is just the term $10C$, which we know is 2, as that is the constant in the original numerator.

This results in five equations in five unknowns, A, B, C, D and E:

$$A + D = 1; \quad 2A + B + E = 0; \quad 10A + 2B + C = 0;$$
$$10B + 2C = 0; \quad 10C = 2.$$

In this case, we can quickly determine C, then back-substitute to obtain the other values. From the fifth equation, we have $C = 1/5$. Now knowing C, in the fourth equation, we have $10B + 2(1/5) = 0$, so that $B = -2/50$, or $-1/25$ after simplification.

Now knowing B and C, the value of A can be determined from the third equation. We have $10A + 2(-1/25) + 1/5 = 0$, so that $A = -3/250$. By the second equation, knowing A and B, we find that $E = 8/125$, and from the first equation, we have $D = 1 + 3/250 = 253/250$.

Thus, the partial fraction decomposition is complete:

$$\frac{x^4 + 2}{x^3(x^2 + 2x + 10)} = -\frac{3}{250x} - \frac{1}{25x^2} + \frac{1}{5x^3} + \frac{253x + 16}{250(x^2 + 2x + 10)}.$$

These last two examples correspond to Examples 19.1 and 19.2. The partial fraction decomposition is the most time-consuming step of inverting a Laplace Transform. The methods shown in these examples illustrate that there are ways to shorten the workload by being focused in strategic areas.

Technology Option

Clearly, there are ways to greatly shorten the amount of work needed by using technology. The TI-83/84 calculators have a built-in RREF (reduced-row echelon form) subroutine that quickly determines the unique solution of an n-variable, n-equation system (assuming a solution exists). This is helpful for systems of three or more unknowns in three or more equations. In this previous example, we would write the five equations, "zero-filling" so that the unknowns A through E are stacked visually:

$$1A + 0B + 0C + 1D + 0E = 1$$
$$2A + 1B + 0C + 0D + 1E = 0$$
$$10A + 2B + 1C + 0D + 0E = 0$$
$$0A + 10B + 2C + 0D + 0E = 0$$
$$0A + 0B + 10C + 0D + 0E = 2$$

This is written as an augmented system, meaning the constants to the right are included as a column. This is a 5 × 6 matrix:

$$\begin{bmatrix} 1 & 0 & 0 & 1 & 0 & 1 \\ 2 & 1 & 0 & 0 & 1 & 0 \\ 10 & 2 & 1 & 0 & 0 & 0 \\ 0 & 10 & 2 & 0 & 0 & 0 \\ 0 & 0 & 10 & 0 & 0 & 2 \end{bmatrix}.$$

Using a TI-83/84 calculator, follow these steps:

1. Press **MATRX** and tab to the third column, **EDIT**, then select **[A]** and press **ENTER**.
2. Enter the size, 5 by 6. Then enter the value in each cell, pressing **ENTER** after each one. A common error is to forget to press **ENTER** after the very last entry. Once done, this is matrix **[A]**, saved in memory.
3. Press **MATRX** again and in the second column **MATH**, scroll down and select **b:rref(** and press **ENTER**. Then press **MATRX** again, and in the **NAMES** column, select **[A]** (it will be default) and press **ENTER**. Your calculator screen will read **rref([A]**. You can type in the right parenthesis, or simply press **ENTER**. The result is

$$\begin{bmatrix} 1 & 0 & 0 & 0 & 0 & -.012 \\ 0 & 1 & 0 & 0 & 0 & -.04 \\ 0 & 0 & 1 & 0 & 0 & .2 \\ 0 & 0 & 0 & 1 & 0 & 1.012 \\ 0 & 0 & 0 & 0 & 1 & .064 \end{bmatrix}.$$

4. To convert into fraction form, select **MATH**, then ▶**Frac**, and press **ENTER**. The result is

$$\begin{bmatrix} 1 & 0 & 0 & 0 & 0 & -3/250 \\ 0 & 1 & 0 & 0 & 0 & -1/25 \\ 0 & 0 & 1 & 0 & 0 & 1/5 \\ 0 & 0 & 0 & 1 & 0 & 253/250 \\ 0 & 0 & 0 & 0 & 1 & 8/125 \end{bmatrix}.$$

This form of a matrix is called reduced-row echelon form, abbreviated RREF. The 1's in the diagonal form a "stair-step" effect (hence, the name "echelon"), and the 1's, in order, correspond to the unknowns A through E. The first row indicates that $A = -3/250$, the second row indicates that $B = -1/25$, and so on.

Variation of Parameters

Variation of Parameters expands on the Method of Undetermined Coefficients primarily for second-degree linear non-homogeneous differential equations, but allows for variable coefficients and is more general, solving forms that other common solution processes cannot.

Given a second-order linear non-homogeneous ODE of the form

$$y'' + f(x)y' + g(x)y = h(x),$$

we first find its homogeneous solution,

$$y_h = C_1 y_1(x) + C_2 y_2(x).$$

The component solution functions $y_1(x)$ and $y_2(x)$ are linearly independent, confirmed by showing their Wronskian is not 0:

$$W(y_1, y_2) \neq 0.$$

It is important to calculate the Wronskian as it will be part of the calculation in a subsequent step. Note that when we write $W(y_1, y_2)$, and since y_1 and y_2 are both functions of x, that for the purposes of the integrals to follow, we write $W(x)$ to emphasize that the integrals are all functions of x (or the independent variable).

A particular solution to the non-homogeneous equation is given by

$$y_p = -y_1(x) \int \frac{y_2(x)h(x)}{W(x)} dx + y_2(x) \int \frac{y_1(x)h(x)}{W(x)} dx.$$

The general solution is the sum of the homogeneous solution and the particular solution,

$$y = y_h + y_p.$$

The principal drawback of this method of solution are the two integrals, which often cannot be antidifferentiated into common functions.

Example: Find the general solution of $y'' + 2y' - 15y = 6e^{3x}$. (This is Example 15.2)

Solution: The homogeneous solution is $y_h = C_1 e^{-5x} + C_2 e^{3x}$.

The Wronskian is

$$W(e^{-5x}, e^{3x}) = \det \begin{bmatrix} e^{-5x} & e^{3x} \\ -5e^{-5x} & 3e^{3x} \end{bmatrix} = 8e^{-2x}.$$

A particular solution is

$$y_p = -e^{-5x} \int \frac{(e^{3x})(6e^{3x})}{8e^{-2x}} dx + e^{3x} \int \frac{(e^{-5x})(6e^{3x})}{8e^{-2x}} dx.$$

Simplify:

$$y_p = -e^{-5x} \int \frac{6e^{6x}}{8e^{-2x}} dx + e^{3x} \int \frac{6e^{-2x}}{8e^{-2x}} dx.$$

$$= -\frac{3}{4} e^{-5x} \int e^{8x} dx + \frac{3}{4} e^{3x} \int dx.$$

Now, integrate:

$$y_p = -\frac{3}{4} e^{-5x} \left(\frac{1}{8} e^{8x} \right) + \frac{3}{4} e^{3x} (x).$$

Any constants of integration can be set to 0 (or ignored ... same thing).

Simplify:

$$y_p = -\frac{3}{32} e^{-5x} + \frac{3}{4} x e^{3x}.$$

The initial form of the general solution is

$$y = C_1 e^{-5x} + C_2 e^{3x} - \frac{3}{32} e^{-5x} + \frac{3}{4} x e^{3x}.$$

The two terms containing e^{-5x} are alike and are combined together, the coefficient $-\frac{3}{32}$ absorbed by the generic constant C_1. The general solution is

$$y = C_1 e^{-5x} + C_2 e^{3x} + \frac{3}{4} x e^{3x}.$$

This is the same solution from Example 15.2. In the Method of Undetermined Coefficients, we had to make an educated guess of the form for the particular solution before proceeding. In this method, the need to make such a guess is not necessary but the basis functions from the homogeneous solution must be known first as they are necessary to set up and evaluate the two integrals.

Example: Find the general solution of $x^2 y'' + 4xy' + 2y = e^x$.

Solution: This is the differential equation from Example 17.1. Its homogeneous solution is $y_h = C_1 x^{-1} + C_2 x^{-2}$, and the Wronskian is $W = x^{-4}$. In order to use the Variation of Parameters formula, the leading coefficient must be 1, so the x^2 is divided through, giving $y'' + 4x^{-1} y' + 2x^{-2} y = x^{-2} e^x$. The homogeneous solutions and Wronskian do not change.

The particular solution is given by

$$y_p = -x^{-1} \int \frac{(x^{-2})(x^{-2} e^x)}{x^{-4}} dx + x^{-2} \int \frac{(x^{-1})(x^{-2} e^x)}{x^{-4}} dx$$

$$= -x^{-1} \int e^x \, dx + x^{-2} \int x e^x \, dx.$$

$$= -x^{-1} e^x + x^{-2}(x e^x - e^x)$$

$$= -x^{-1} e^x + x^{-1} e^x + x^{-2} e^x$$

$$= x^{-2} e^x.$$

In the integration step, the second integral is determined by integration-by-parts. The general solution is

$$y = C_1 x^{-1} + C_2 x^{-2} + x^{-2} e^x, \quad x \neq 0.$$

Example: Find the general solution of $y'' + 5y' - 14y = \sin 3x$.

Solution: The homogeneous solution is $y_h = C_1 e^{2x} + C_2 e^{-7x}$ and the Wronskian is $W = -9e^{-5x}$. The particular solution is

$$y_p = -e^{2x} \int \frac{(e^{-7x})(\sin 3x)}{-9e^{-5x}} dx + e^{-7x} \int \frac{(e^{2x})(\sin 3x)}{-9e^{-5x}} dx$$

$$= \frac{1}{9} e^{2x} \int e^{-2x} \sin 3x \, dx - \frac{1}{9} e^{-7x} \int e^{7x} \sin 3x \, dx.$$

Both integrals are solved by integration-by-parts or using a table of integrals. They are:

$$\int e^{-2x} \sin 3x \, dx = -\frac{1}{13} e^{-2x}(2 \sin 3x + 3 \cos 3x),$$

$$\int e^{7x} \sin 3x \, dx = \frac{1}{58} e^{7x}(7 \sin 3x - 3 \cos 3x).$$

Assembling the expressions, the particular solution is

$$y_p = \frac{1}{9} e^{2x} \left(-\frac{1}{13} e^{-2x}(2 \sin 3x + 3 \cos 3x) \right)$$

$$- \frac{1}{9} e^{-7x} \left(\frac{1}{58} e^{7x}(7 \sin 3x - 3 \cos 3x) \right)$$

$$= -\frac{1}{117}(2 \sin 3x + 3 \cos 3x) - \frac{1}{522}(7 \sin 3x - 3 \cos 3x)$$

$$= \left(-\frac{2}{117} - \frac{7}{522} \right) \sin 3x + \left(\frac{3}{522} - \frac{3}{117} \right) \cos 3x$$

$$= -\frac{23}{754} \sin 3x - \frac{15}{754} \cos 3x.$$

The general solution is

$$y = C_1 e^{2x} + C_2 e^{-7x} - \frac{23}{754} \sin 3x - \frac{15}{754} \cos 3x.$$

This method works insofar that the integrands can be antidifferentiated. For example, consider the previous example, but with ln (x) in place of sin (3x). The particular solution can be set up:

$$y_p = \frac{1}{9} e^{2x} \int e^{-2x} \ln(x) \, dx - \frac{1}{9} e^{-7x} \int e^{7x} \ln(x) \, dx.$$

Neither integral can be evaluated into common functions. This is as far as this method will take us on this example. The solution shown above can only be kept as is.

Some of the problems posed in Section 15 (Undetermined Coefficients) and Section 17 (Cauchy-Euler Equations), if the forcing function is not 0, can be solved using Variation of Parameters, as demonstrated here. You should try both methods out of curiosity and judge which one is preferable.

Exact Equations

This section requires some knowledge of multivariable functions and partial differentiation. Thus, it may not be taught in many introductory differential equations courses that usually do not require multivariable calculus as a prerequisite.

An *explicit function* has one of its variables isolated. For example, $y = x^2 + 2x$ is an explicit function with y isolated. If the terms are collected to one side, such as $y - x^2 - 2x = 0$, we have an *implicit function* set equal to a constant. But remember, y is still some function of x, in that if a value for x is chosen, then y's value is dependent on x. When possible, we would solve for y. But sometimes, that's not possible. For example, consider

$$x^3 y^4 + 2xy - 3x^2 = 0.$$

This is an implicit function set equal to 0. There is no way to algebraically isolate y. Sometimes, the implicit form is cleaner than the explicit form. For example, a circle is described implicitly by $x^2 + y^2 = r^2$, or explicitly by $y = \pm\sqrt{r^2 - x^2}$. Which version do you like?

In multivariable calculus, the above expression could be defined as a multivariable function:

$$f(x, y) = x^3 y^4 + 2xy - 3x^2.$$

Normally, variables x and y are independently chosen. But when the equation is set equal to a constant, $f(x, y) = x^3 y^4 + 2xy - 3x^2 = C$, the variables are not independent of one another. Choosing a value for one will "force" the other to take on a certain value(s) in order for the statement to be true. If we keep in mind that y itself is related to x by the expression $y(x)$, then the function above can be written

$$f(x, y(x)) = x^3 [y(x)]^4 + 2x[y(x)] - 3x^2 = C.$$

The derivative of the above function with respect to x is

$$\frac{\partial f}{\partial x}\left(\frac{dx}{dx}\right) + \frac{\partial f}{\partial y}\left(\frac{dy}{dx}\right),$$

Where $\frac{\partial f}{\partial x}$ is the partial derivative of f with respect to x, and $\frac{\partial f}{\partial y}$ is the partial derivative of f with respect to y. This is the Chain Rule form of the derivative for multi-variable functions. For convenience, we often write f_x for $\frac{\partial f}{\partial x}$ and f_y for $\frac{\partial f}{\partial y}$.

Remember, this was all set to 0 many steps previously, and noting that $\frac{dx}{dx} = 1$, we have

$$\frac{\partial f}{\partial x} + \frac{\partial f}{\partial y}\left(\frac{dy}{dx}\right) = 0.$$

A typical problem with this structure might be

$$2xy + x^2 \frac{dy}{dx} = 0.$$

There may exist some function $f(x, y)$ such that $f_x = 2xy$ and $f_y = x^2$. If so, then $f(x, y) = C$ represents the solution of the above differential equation. The above differential equation is called an **exact equation**, and it handles cases where the solution function is usually written implicitly. An exact equation has the form

$$M(x, y) + N(x, y)\frac{dy}{dx} = 0,$$

where $M(x, y) = f_x(x, y)$ and $N(x, y) = f_y(x, y)$, assuming $f(x, y)$ exists. How do we know if such a function (solution) exists? And how do we find it?

Assuming a solution function $f(x, y)$ exists, and is twice differentiable, then Clairaut's Theorem states that $f_{xy} = f_{yx}$, or in the above case, we check to see whether $M_y = N_x$. If true, then there exists a solution function $f(x, y)$, and if false, there does not. In calculus, such a function is called a *potential* function.

Example: Show whether a function $f(x, y)$ exists that solves $2xy + x^2 \frac{dy}{dx} = 0$, and if so, find this function.

Solution: We have $M(x, y) = 2xy$ and $N(x, y) = x^2$. Therefore, $M_y = 2x$ and $N_x = 2x$. Since $M_y = N_x$, there does exist a function written in implicit form, $f(x, y) = c$, that solves this differential equation.

We know that $f_x = M$ and that $f_y = N$. Thus, we integrate M with respect to x and N with respect to y and study the results:

$$\int \underbrace{2xy}_{M}\, dx = x^2 y + C, \quad \text{and} \quad \int \underbrace{x^2}_{N}\, dy = x^2 y + C.$$

It appears that $f(x, y) = x^2 y$ is the solution. To be sure, verify that $f_x = M$ and that $f_y = N$. Thus, the solution to the differential equation is

$$x^2 y = c.$$

What happens if we tried to solve this differential equation using separation of variables? We have

$$x^2 \frac{dy}{dx} = -2xy$$

$$x \frac{dy}{dx} = -2y, \quad x \neq 0$$

$$-\frac{dy}{2y} = \frac{dx}{x}$$

$$-\int \frac{dy}{2y} = \int \frac{dx}{x}$$

$$-\frac{1}{2} \ln(y) = \ln(x) + C$$

$$\ln(y^{-1/2}) = \ln(x) + C$$

$$y^{-1/2} = Cx \quad \text{(taking base } e \text{ of both sides)}$$

$$y = Cx^{-2} = \frac{C}{x^2}.$$

This is a solution. It checks:

$$2x \left(\overbrace{\frac{C}{x^2}}^{y}\right) + x^2 \left(\overbrace{-\frac{2C}{x^3}}^{y'}\right) = \frac{2C}{x} - \frac{2C}{x} = 0.$$

Note that $y = \frac{C}{x^2}$ is the same as writing $x^2 y = C$. In this case, the differential equation was exact, and it could be solved using its method, or could be solved in another way (in this example, separation of variables). This is a case where it was not difficult to isolate y. Not all such equations are solvable via separation of variables.

If it is determined that a potential function exists, finding it may take some clever insight. The following example illustrates a heuristic way to find the potential function.

183

Example: Given $(3x^2y^4 + 2y - 6x) + (4x^3y^3 + 2x)\frac{dy}{dx}$. Show whether a potential function that solves this differential equation exists and if so, find it.

Solution: We have $M(x,y) = 3x^2y^4 + 2y - 6x$ and $N(x,y) = 4x^3y^3 + 2x$. We show that $M_y = N_x$:

$$M_y = 12x^2y^3 + 2, \quad \text{and} \quad N_x = 12x^2y^3 + 2.$$

Since $M_y = N_x$, a potential function $f(x,y)$ exists where $f_x = M$ and $f_y = N$. To find it, antidifferentiate M with respect to x and N with respect to y. For the time being, any constants of integration can be ignored.

$$\int M\,dx = \int (3x^2y^4 + 2y - 6x)\,dx = 3\left(\frac{1}{3}x^3\right)y^4 + 2xy - 3x^2$$
$$= x^3y^4 + 2xy - 3x^2,$$

$$\int N\,dy = \int (4x^3y^3 + 2x)\,dy = 4x^3\left(\frac{1}{4}y^4\right) + 2xy = x^3y^4 + 2xy.$$

Note that x^3y^4 and $2xy$ both appear in each's antiderivative. Thus, it is reasonable to infer that the potential function contains these two terms. Also, note that $-3x^2$ appears in the antiderivative of M. But it lacks a y, so that if differentiated with respect to y, it would result in 0. Thus, the potential function is (probably):

$$f(x,y) = x^3y^4 + 2xy - 3x^2.$$

Check that this is true by confirming that $f_x = M$ and $f_y = N$. If so, then the solution in implicit form is

$$x^3y^4 + 2xy - 3x^2 = C.$$

Note two things: (1) the original differential equation is not separable, and (2) it would be impossible to isolate y or x in the solution. It can only be written in implicit form.

This method of finding the potential function is not the formal method, but it works for almost all cases. Once the antiderivatives $\int M\,dx$ and $\int N\,dy$ have been determined, the potential function is (probably) the union of such terms. Terms that appear in both antiderivatives need not be written twice when forming the potential function. If using this method, it is imperative that you check that $f_x = M$ and $f_y = N$.

Example: Given $(3x^5y - 7xy) + (x^3y^2 + x)\frac{dy}{dx}$. Show whether a potential function that solves this differential equation exists and if so, find it.

Solution: We determine whether $M_y = N_x$ is true:

$$M_y = 3x^5 - 7x, \quad \text{and} \quad N_x = 3x^2y^2 + 1.$$

Since $M_y \neq N_x$, this differential equation is not exact, and no potential function exists. It is not separable either. A solution exists, but no easy method exists to find it.

Series Solutions

A *series* is a sum of terms. If the terms proceed forever, it is called an *infinite series*. A *power series* is an infinite series in which each term contains the independent variable x raised to a non-negative integer power. Summation notation is used to represent a series.

A power series has the form

$$\sum_{n=0}^{\infty} a_n x^n = a_0 + a_1 x + a_2 x^2 + a_3 x^3 + a_4 x^4 + a_5 x^5 + \cdots.$$

The n is the index. Certain common functions can be represented by power series centered at $x = 0$. Three common ones are the power series for e^x, $\cos(x)$ and $\sin(x)$. They are

$$e^x = 1 + x + \frac{1}{2!}x^2 + \frac{1}{3!}x^3 + \frac{1}{4!}x^4 + \frac{1}{5!}x^5 + \frac{1}{6!}x^6 + \frac{1}{7!}x^7 + \frac{1}{8!}x^8 + \cdots,$$

$$\cos x = 1 - \frac{1}{2!}x^2 + \frac{1}{4!}x^4 - \frac{1}{6!}x^6 + \frac{1}{8!}x^8 - \cdots,$$

$$\sin x = x - \frac{1}{3!}x^3 + \frac{1}{5!}x^5 - \frac{1}{7!}x^7 + \frac{1}{9!}x^9 - \cdots.$$

These are also called Maclaurin series and are used commonly in calculus to approximate such functions using polynomials of ever-increasing degree. Let's solve a couple differential equations in which the solution is already known, and see how well this method works:

Example: Use power series to solve $y' = y$ centered at $x = 0$. We already know that the general solution is $y = Ce^x$.

Solution: We assume the solution has the form

$$y = a_0 + a_1 x + a_2 x^2 + a_3 x^3 + a_4 x^4 + a_5 x^5 + \cdots = \sum_{n=0}^{\infty} a_n x^n.$$

Differentiating gives

$$y' = a_1 + 2a_2 x + 3a_3 x^2 + 4a_4 x^3 + 5a_5 x^4 + \cdots = \sum_{n=1}^{\infty} n a_n x^{n-1}.$$

We want both series to begin at the same index n. Since the series for y' starts at $n = 1$, let's rewrite the series for y so that it starts at $n = 1$ as well. Replace n with $n - 1$, we get

$$\sum_{n=0}^{\infty} a_n x^n = \sum_{n-1=0}^{\infty} a_{n-1} x^{n-1} = \sum_{n=1}^{\infty} a_{n-1} x^{n-1}.$$

Since $y' = y$ can be rewritten $y' - y = 0$, substitute in the series expressions:

$$\underbrace{\sum_{n=1}^{\infty} n a_n x^{n-1}}_{y'} - \underbrace{\left(\sum_{n=1}^{\infty} a_{n-1} x^{n-1} \right)}_{y} = 0.$$

Since both summations start at $n = 1$ and since the powers of x match, combine now into a single summation:

$$\sum_{n=1}^{\infty} [n a_n - a_{n-1}] x^{n-1} = 0.$$

For this to be the solution, all coefficients of x must be zero. Thus, we have

$$n a_n - a_{n-1} = 0, \quad n = 1, 2, 3, \ldots \; .$$

Isolate a_n:

$$a_n = \frac{1}{n} a_{n-1}.$$

This is a recursive relationship, in which each coefficient is defined in terms of the one preceding it. We write out a few to detect a pattern.

$$n = 1: \quad a_1 = \frac{1}{1}a_0 = \frac{1}{1!}a_0$$

$$n = 2: \quad a_2 = \frac{1}{2}a_1 = \frac{1}{2}\cdot\frac{1}{1}a_0 = \frac{1}{2!}a_0$$

$$n = 3: \quad a_3 = \frac{1}{3}a_2 = \frac{1}{3}\cdot\frac{1}{2}\cdot\frac{1}{1}a_0 = \frac{1}{3!}a_0$$

$$n = 4: \quad a_4 = \frac{1}{4}a_3 = \frac{1}{4}\cdot\frac{1}{3}\cdot\frac{1}{2}\cdot\frac{1}{1}a_0 = \frac{1}{4!}a_0$$

$$n = 5: \quad a_5 = \frac{1}{5}a_4 = \frac{1}{5}\cdot\frac{1}{4}\cdot\frac{1}{3}\cdot\frac{1}{2}\cdot\frac{1}{1}a_0 = \frac{1}{5!}a_0$$

This generalizes to

$$a_n = \frac{1}{n!}a_0, \quad n = 1, 2, 3, 4, \ldots .$$

The original assumption was that the solution had the form

$$y = a_0 + a_1 x + a_2 x^2 + a_3 x^3 + a_4 x^4 + a_5 x^5 + \cdots = \sum_{n=0}^{\infty} a_n x^n.$$

Replacing a_1, a_2, a_3, and so on, with their equivalents in terms of a_0 gives

$$y = a_0 + \frac{1}{1!}a_0 x + \frac{1}{2!}a_0 x^2 + \frac{1}{3!}a_0 x^3 + \frac{1}{4!}a_0 x^4 + \frac{1}{5!}a_0 x^5 + \cdots = \sum_{n=0}^{\infty} \left(\frac{a_0}{n!}\right) x^n.$$

The a_0 factors to the front, leaving

$$a_0 \sum_{n=0}^{\infty} \frac{x^n}{n!}.$$

The series above is the Maclaurin series for e^x. Thus, the solution is $y = a_0 e^x$, where the generic constant C is represented by a_0.

Example: Find a solution in series form for $y'' - xy = 0$, centered about $x = 0$.

Solution: As before, we assume the series has the form

$$y = a_0 + a_1 x + a_2 x^2 + a_3 x^3 + a_4 x^4 + a_5 x^5 + \cdots = \sum_{n=0}^{\infty} a_n x^n.$$

Differentiating twice gives

$$y' = \sum_{n=1}^{\infty} n a_n x^{n-1} \quad \text{and} \quad y'' = \sum_{n=2}^{\infty} n(n-1) a_n x^{n-2}.$$

Substituting into the differential equation gives

$$\sum_{n=2}^{\infty} n(n-1) a_n x^{n-2} - x \sum_{n=0}^{\infty} a_n x^n = 0.$$

Distribute the x through the last series:

$$\sum_{n=2}^{\infty} n(n-1) a_n x^{n-2} - \sum_{n=0}^{\infty} a_n x^{n+1} = 0.$$

We need to do two things: 1) adjust the indices to that each series has x to the same power, and 2) further adjust the indices so that each series starts at the same lower index.

For the first action, write $n + 2$ for n in the first series and $n - 1$ for n in the last series. This will make all the powers of x the same (n in this case).

$$\sum_{n=0}^{\infty} (n+2)(n+1) a_{n+2} x^n - \sum_{n=1}^{\infty} a_{n-1} x^n = 0.$$

From the first series, "pull out" the term for $n = 0$, which is $2a_2$. This resets the first series' lower index to 1 but does not alter the summand form within it. We have

$$2a_2 + \sum_{n=1}^{\infty} (n+2)(n+1) a_{n+2} x^n - \sum_{n=1}^{\infty} a_{n-1} x^n = 0.$$

Now the two series can be combined into one. The x^n is factored so that the coefficient form is better seen:

$$2a_2 + \sum_{n=1}^{\infty} [(n+2)(n+1)a_{n+2} - a_{n-1}]x^n = 0.$$

Each coefficient must be 0. This means that

$$(n+2)(n+1)a_{n+2} - a_{n-1} = 0,$$

and solving for a_{n+2}, we get

$$a_{n+2} = \frac{1}{(n+2)(n+1)} a_{n-1}.$$

We'll worry about the a_0, a_1 and a_2 terms in a moment.

For the indices of n, we evaluate a_{n+2}:

$$n = 1: \quad a_3 = \frac{1}{3 \cdot 2} a_0,$$

$$n = 2: \quad a_4 = \frac{1}{4 \cdot 3} a_1,$$

$$n = 3: \quad a_5 = \frac{1}{5 \cdot 4} a_2,$$

$$n = 4: \quad a_6 = \frac{1}{6 \cdot 5} a_3 = \frac{1}{6 \cdot 5 \cdot 3 \cdot 2} a_0,$$

$$n = 5: \quad a_7 = \frac{1}{7 \cdot 6} a_4 = \frac{1}{7 \cdot 6 \cdot 4 \cdot 3} a_1,$$

$$n = 6: \quad a_8 = \frac{1}{8 \cdot 7} a_5 = \frac{1}{8 \cdot 7 \cdot 5 \cdot 4} a_2,$$

$$n = 7: \quad a_9 = \frac{1}{9 \cdot 8} a_6 = \frac{1}{9 \cdot 8 \cdot 6 \cdot 5 \cdot 3 \cdot 2} a_0,$$

$$n = 8: \quad a_{10} = \frac{1}{10 \cdot 9} a_7 = \frac{1}{10 \cdot 9 \cdot 7 \cdot 6 \cdot 4 \cdot 3} a_1,$$

$$n = 9: \quad a_{11} = \frac{1}{11 \cdot 10} a_8 = \frac{1}{11 \cdot 10 \cdot 8 \cdot 7 \cdot 5 \cdot 4} a_2$$

and so on.

Note that the coefficients a_n, for $n = 3, 6, 9, 12$, etc., are stated in terms of a_0. Similarly, the coefficients for $n = 1, 4, 7, 10$, etc., are stated in terms of a_1, and for $n = 2, 5, 8, 11$, etc., are stated in terms of a_2. Grouping them, we have:

$$a_0 + a_3 x^3 + a_6 x^6 + a_9 x^9 + \cdots$$
$$= a_0 + \frac{a_0}{3 \cdot 2} x^3 + \frac{a_0}{6 \cdot 5 \cdot 3 \cdot 2} x^6 + \frac{a_0}{9 \cdot 8 \cdot 6 \cdot 5 \cdot 3 \cdot 2} x^9 + \cdots,$$

$$a_1 x + a_4 x^4 + a_7 x^7 + a_{10} x^{10} + \cdots$$
$$= a_1 x + \frac{a_1}{4 \cdot 3} x^4 + \frac{a_1}{7 \cdot 6 \cdot 4 \cdot 3} x^7 + \frac{a_1}{10 \cdot 9 \cdot 7 \cdot 6 \cdot 4 \cdot 3} x^{10} + \cdots,$$

$$a_2 x^2 + a_5 x^5 + a_8 x^8 + a_{11} x^{11} + \cdots$$
$$= a_2 x^2 + \frac{a_2}{5 \cdot 4} x^5 + \frac{a_2}{8 \cdot 7 \cdot 5 \cdot 4} x^8 + \frac{a_2}{11 \cdot 10 \cdot 8 \cdot 7 \cdot 5 \cdot 4} x^{11} + \cdots.$$

We had pulled out the term $2a_2$ from the original series a few steps back. This term must be 0, so that causes terms a_5, a_8, a_{11} and so on, to vanish.

The series solution starts to take shape:

$$y = a_0 \left(1 + \frac{x^3}{3 \cdot 2} + \frac{x^6}{6 \cdot 5 \cdot 3 \cdot 2} + \frac{x^9}{9 \cdot 8 \cdot 6 \cdot 5 \cdot 3 \cdot 2} + \cdots \right)$$
$$+ a_1 \left(x + \frac{x^4}{4 \cdot 3} + \frac{x^7}{7 \cdot 6 \cdot 4 \cdot 3} + \frac{x^{10}}{10 \cdot 9 \cdot 7 \cdot 6 \cdot 4 \cdot 3} + \cdots \right).$$

The values for a_0 and a_1 would be found by evaluating at any initial conditions at $x = 0$. That is, $y(0) = a_0$ and $y'(0) = a_1$.

A pattern in the coefficients is clear enough to find further terms. It is up to you if you want to generalize as a formula or leave it as is.

The solution to the differential equation $y'' - xy = 0$, which we just found in series form, consists of two series. The series associated with a_0 is called Airy's Function, and the series associated with a_1 is called the Related Airy's Function (or the Bi-function) These are linearly independent functions and form the general solution (as shown above) of the differential equation. These functions are commonly seen in physics and engineering applications.

Example: Find a solution in series form for $y'' - y' - xy = 0$, centered about $x = 0$.

Solution: We have

$$y = \sum_{n=0}^{\infty} a_n x^n, \quad y' = \sum_{n=1}^{\infty} n a_n x^{n-1} \quad \text{and} \quad y'' = \sum_{n=2}^{\infty} n(n-1) a_n x^{n-2}.$$

Substituting into the differential equation gives

$$\sum_{n=2}^{\infty} n(n-1) a_n x^{n-2} - \sum_{n=1}^{\infty} n a_n x^{n-1} - x \sum_{n=0}^{\infty} a_n x^n = 0.$$

Distribute the x through the last series:

$$\sum_{n=2}^{\infty} n(n-1) a_n x^{n-2} - \sum_{n=1}^{\infty} n a_n x^{n-1} - \sum_{n=0}^{\infty} a_n x^{n+1} = 0.$$

Write $n + 2$ for n in the first series, $n + 1$ for n in the middle series, and $n - 1$ for n in the last series. This will make all the powers of x the same (n in this case).

$$\sum_{n=0}^{\infty} (n+2)(n+1) a_{n+2} x^n - \sum_{n=0}^{\infty} (n+1) a_{n+1} x^n - \sum_{n=1}^{\infty} a_{n-1} x^n = 0.$$

Next, we "pull out" the term for $n = 0$ from the first two series, which then resets the series' lower bound to 1. Note that this maneuver does not change the general form within the series:

$$2a_2 - a_1 + \sum_{n=1}^{\infty} (n+2)(n+1) a_{n+2} x^n - \sum_{n=1}^{\infty} (n+1) a_{n+1} x^n - \sum_{n=1}^{\infty} a_{n-1} x^n = 0.$$

Now that all three series have the same power of x and start from the same lower index, they can be combined into one series. Factor the x^n in the same step:

$$-a_1 + 2a_2 + \sum_{n=1}^{\infty} [(n+2)(n+1) a_{n+2} - (n+1) a_{n+1} - a_{n-1}] x^n = 0.$$

The coefficients are set to 0. We'll worry about the $-a_1 + 2a_2$ in a moment.

$$(n+2)(n+1) a_{n+2} - (n+1) a_{n+1} - a_{n-1} = 0, \quad n = 1, 2, 3, 4, 5, \ldots$$

Solve for a_{n+2}:

$$a_{n+2} = \frac{(n+1)a_{n+1} + a_{n-1}}{(n+2)(n+1)}.$$

This is a three-term recursion, in that the value of a particular term is found by the above formula involving two of the previous terms.

Evaluate for values of n:

$$n = 1: \quad a_3 = \frac{2a_2 + a_0}{3 \cdot 2}$$

$$n = 2: \quad a_4 = \frac{3a_3 + a_1}{4 \cdot 3}$$

$$n = 3: \quad a_5 = \frac{4a_4 + a_2}{5 \cdot 4}$$

$$n = 4: \quad a_6 = \frac{5a_5 + a_3}{6 \cdot 5}$$

$$n = 5: \quad a_7 = \frac{6a_6 + a_4}{7 \cdot 6}$$

The terms a_1 and a_2 are connected by the relationship $-a_1 + 2a_2 = 0$. Thus, whatever is chosen for a_1 forces $a_2 = \frac{1}{2}a_1$. The above calculations are adjusted accordingly:

$$a_3 = \frac{2a_2 + a_0}{3 \cdot 2} = \frac{2\left(\frac{a_1}{2}\right) + a_0}{3 \cdot 2} = \frac{a_1 + a_0}{3!},$$

$$a_4 = \frac{3a_3 + a_1}{4 \cdot 3} = \frac{3\left(\frac{a_1 + a_0}{3!}\right) + a_1}{4 \cdot 3} = \frac{3a_1 + a_0}{4!},$$

$$a_5 = \frac{4a_4 + a_2}{5 \cdot 4} = \frac{4\left(\frac{3a_1 + a_0}{4!}\right) + \left(\frac{a_1}{2}\right)}{5 \cdot 4} = \frac{6a_1 + a_0}{5!},$$

$$a_6 = \frac{5a_5 + a_3}{6 \cdot 5} = \frac{5\left(\frac{6a_1 + a_0}{5!}\right) + \left(\frac{a_1 + a_0}{3!}\right)}{6 \cdot 5} = \frac{10a_1 + 5a_0}{6!},$$

$$a_7 = \frac{6a_6 + a_4}{7 \cdot 6} = \frac{6\left(\frac{10a_1 + 5a_0}{6!}\right) + \left(\frac{3a_1 + a_0}{4!}\right)}{7 \cdot 6} = \frac{25a_1 + 10a_0}{7!},$$

$$a_8 = \frac{7a_7 + a_5}{8 \cdot 7} = \frac{7\left(\frac{25a_1 + 10a_0}{7!}\right) + \left(\frac{6a_1 + a_0}{5!}\right)}{8 \cdot 7} = \frac{61a_1 + 16a_0}{8!}, \quad \text{etc.}$$

Note that all terms a_n for $n \geq 2$ are defined in terms of a_0 and a_1. Thus, the solution to $y'' - y' - xy = 0$ can be written as two series shown below, one dependent on a_0 and the other on a_1. These becomes the two generic constants of the general solution.

$$y = a_0 \left(1 + \frac{1}{3!}x^3 + \frac{1}{4!}x^4 + \frac{1}{5!}x^5 + \frac{5}{6!}x^6 + \frac{10}{7!}x^7 + \frac{16}{8!}x^8 + \cdots \right)$$
$$+ a_1 \left(x + \frac{1}{2!}x^2 + \frac{1}{3!}x^3 + \frac{3}{4!}x^4 + \frac{6}{5!}x^5 + \frac{10}{6!}x^6 + \frac{25}{7!}x^7 + \frac{61}{8!}x^8 + \cdots \right)$$

These two series form the solution basis and are linearly independent. The Wronskian will always result in an expression with the term $a_0 a_1$ in it (along with a string of terms of various non-zero powers of x). As long as $a_0 \neq 0$ and $a_1 \neq 0$, the Wronskian will always be non-zero. Another way to see this is that if the two series expressions were linearly dependent, one would be a non-zero scalar multiple of the other. This cannot be the case here. For example, the first series contains a constant (the 1) whereas the second does not. There is no way to multiply the first series by a non-zero constant so that the constant vanishes.

Since this differential equation was solved by a series centered at $x = 0$, this means that any initial conditions would have to be at $x = 0$. That is, we would specify the initial conditions as $y(0) = h$ and $y'(0) = k$. What happens in this case?

At $x = 0$, the entire second series vanishes, and all terms except the leading 1 vanish in the first series. As a result, $a_0 = h$. Now, taking the derivative of y, the entire first series consists of terms with non-zero powers of x, and the second series now has a constant (the lone x differentiates to 1). Upon substitution of $x = 0$, the entire first series vanishes and all terms in the second series vanish except for the leading 1. This forces $a_1 = k$. Thus, the initial conditions can be summarized as $y(0) = a_0$ and $y'(0) = a_1$.

If both $a_0 = 0$ and $a_1 = 0$, then $y = 0$ and we have a trivial (non-interesting) solution to the differential equation.

If either $a_0 = 0$ or $a_1 = 0$, then the solution reduces to just one series.

If both a_0 and a_1 are non-zero, then both series are present, the Wronskian is not 0, and the series are linearly independent.

Lastly, "centered" at $x = 0$ means that the series solution we have found is valid for x-values close to 0. The series is used to approximate solution points for these small values of x. More precision can be found by including more terms.

A series can be used to numerically approximate solutions when a closed-form solution cannot be found.

Example: In Example 6.4, we saw that the solution of $y' + 2xy = 1$ is

$$y = \frac{\int e^{x^2} dx + C}{e^{x^2}}.$$

Solution: It is impossible to resolve $\int e^{x^2} dx$ into common functions. In other words, there is a solution, but it cannot be represented in the familiar functions we have available. Instead, we can use a series expansion for e^{x^2} and approximate solutions that way. Assuming the solution is centered at $x = 0$ (In Example 9.2, with an initial condition of $y(0) = 1$, it was), we use the Maclaurin series for e^x, which is

$$e^x = 1 + x + \frac{1}{2!}x^2 + \frac{1}{3!}x^3 + \frac{1}{4!}x^4 + \frac{1}{5!}x^5 + \frac{1}{6!}x^6 + \frac{1}{7!}x^7 + \frac{1}{8!}x^8 + \cdots.$$

Write x^2 in place of x and simplify:

$$e^{x^2} = 1 + (x^2) + \frac{1}{2!}(x^2)^2 + \frac{1}{3!}(x^2)^3 + \frac{1}{4!}(x^2)^4 + \frac{1}{5!}(x^2)^5 \ldots$$

$$= 1 + x^2 + \frac{1}{2!}x^4 + \frac{1}{3!}x^6 + \frac{1}{4!}x^8 + \frac{1}{5!}x^{10} + \cdots.$$

Antidifferentiating, we have:

$$\int e^{x^2} dx = \int \left(1 + x^2 + \frac{1}{2!}x^4 + \frac{1}{3!}x^6 + \frac{1}{4!}x^8 + \frac{1}{5!}x^{10} + \cdots\right) dx$$

$$= x + \frac{1}{3}x^3 + \frac{1}{2!}\left(\frac{1}{5}x^5\right) + \frac{1}{3!}\left(\frac{1}{7}x^7\right) + \frac{1}{4!}\left(\frac{1}{9}x^9\right) + \frac{1}{5!}\left(\frac{1}{11}x^{11}\right) + C.$$

Thus, the solution $y = \frac{\int e^{x^2} dx + C}{e^{x^2}}$ can be approximated by

$$y = \frac{x + \frac{1}{3}x^3 + \frac{1}{2!}\left(\frac{1}{5}x^5\right) + \frac{1}{3!}\left(\frac{1}{7}x^7\right) + \frac{1}{4!}\left(\frac{1}{9}x^9\right) + \frac{1}{5!}\left(\frac{1}{11}x^{11}\right) + \cdots + C}{1 + x^2 + \frac{1}{2!}x^4 + \frac{1}{3!}x^6 + \frac{1}{4!}x^8 + \frac{1}{5!}x^{10} + \cdots}.$$

Using the initial condition from Example 9.2, which is (0,1), this makes $C = 1$. The specific solution, with coefficients simplified, is

$$y = \frac{x + \frac{1}{3}x^3 + \frac{1}{10}x^5 + \frac{1}{42}x^7 + \frac{1}{216}x^9 + \frac{1}{1320}x^{11} + \cdots + 1}{1 + x^2 + \frac{1}{2}x^4 + \frac{1}{6}x^6 + \frac{1}{24}x^8 + \frac{1}{120}x^{10} \cdots}.$$

In Example 9.2, we used a numerical method to approximate values of the solution for $x = 0.1, 0.2,$ and 0.3. Let's use this solution in series form to generate y values and compare to what we found in Example 9.2. We'll use the first 6 terms:

$$y = \frac{x + \frac{1}{3}x^3 + \frac{1}{10}x^5 + \frac{1}{42}x^7 + \frac{1}{216}x^9 + \frac{1}{1320}x^{11} + 1}{1 + x^2 + \frac{1}{2}x^4 + \frac{1}{6}x^6 + \frac{1}{24}x^8 + \frac{1}{120}x^{10}}.$$

At $x = 0.1$, we have

$$y = \frac{(0.1) + \frac{1}{3}(0.1)^3 + \frac{1}{10}(0.1)^5 + \frac{1}{42}(0.1)^7 + \frac{1}{216}(0.1)^9 + \frac{1}{1320}(0.1)^{11} + 1}{1 + (0.1)^2 + \frac{1}{2}(0.1)^4 + \frac{1}{6}(0.1)^6 + \frac{1}{24}(0.1)^8 + \frac{1}{120}(0.1)^{10}}$$

$\approx 1.008938\ldots$.

At $x = 0.2$, we have

$$y = \frac{(0.2) + \frac{1}{3}(0.2)^3 + \frac{1}{10}(0.2)^5 + \frac{1}{42}(0.2)^7 + \frac{1}{216}(0.2)^9 + \frac{1}{1320}(0.2)^{11} + 1}{1 + (0.2)^2 + \frac{1}{2}(0.2)^4 + \frac{1}{6}(0.2)^6 + \frac{1}{24}(0.2)^8 + \frac{1}{120}(0.2)^{10}}$$

$\approx 1.155540\ldots$.

At $x = 0.3$, we have

$$y = \frac{(0.3) + \frac{1}{3}(0.3)^3 + \frac{1}{10}(0.3)^5 + \frac{1}{42}(0.3)^7 + \frac{1}{216}(0.3)^9 + \frac{1}{1320}(0.3)^{11} + 1}{1 + (0.3)^2 + \frac{1}{2}(0.3)^4 + \frac{1}{6}(0.3)^6 + \frac{1}{24}(0.3)^8 + \frac{1}{120}(0.3)^{10}}$$

$\approx 1.196562\ldots$.

We have generated the ordered pairs $(0.1, 1.008938\ldots)$, $(0.2, 1.155540\ldots)$ and $(0.3, 1.196562\ldots)$. Using Euler's Method in Example 9.2, we generated ordered pairs $(0.1, 1.1)$, $(0.2, 1.178)$ and $(0.3, 1.23088)$. The approximations generated by the series can be made more precise by using more terms. Keep in mind both methods give approximations of the true values on the solution curve.

Practice

Find the general solution to the following differential equations. If initial conditions are present, find the specific solution.

1. $y' = 10y$
2. $y' + 4xy = 0$
3. $xy' = y^2, \ y(1) = 3$
4. $y' - 4xy = 2x$
5. $y' + 8y = 3x$
6. $y' = x^2\sqrt{y}$
7. $y' + 6xy = x, \ y(0) = 4$
8. $y'' - y' - 20y = 0$
9. $y'' + 4y' - 12y = 0, \ y(0) = 3, y'(0) = 2$
10. $y'' + 50y = 0$
11. $y'' - 14y' + 49y = 0$
12. $y'' + 2y' + 8y = 0$
13. $y^{(3)} - 2y'' - 19y' + 20y = 0$
14. $y^{(3)} - 2y'' + 4y' - 8y = 0$
15. $y^{(3)} - 3y'' - 9y' - 5y = 0$
16. $y^{(4)} + 18y'' + 81y = 0$
17. $y'' + 3y' + 2y = e^{5x}$
18. $y'' - 2y' - 8y = 3e^{-2x}$
19. $y'' + 16y = 3\sin 4x$
20. $y'' + 2y' - 15y = 4x^2 + x$
21. $y' + \frac{1}{x}y = y^2, \ y(1) = 5$
22. $3x^2y'' + 9xy' + 3y = 0$

23. Given that $y_1 = t^3$ is a solution of $t^2y'' + ty' - 9y = 0$. Find another solution y_2, and write the general solution.
24. State the domain of x for the solution of problem 3.
25. State the domain of x for the solution of problem 21.
26. The rate of change of the population of a county is proportional to the square of the county's population. If there were 5000 people in the county in 2016 ($t = 0$), and 6200 people in the county in 2022, find the function that models this county's population and forecast the county's population in 2027.
27. The rate of change of the value of a house is inversely proportional to the cube of its value. If the house was worth $200,000 in 2012 and $225,000 in 2017, find the function that models the value of the house, and forecast the house's value in 2024.

28. A vat contains 2000 gallons of water in which 50 lb of salt is mixed. At one end, a mixture of water containing 2 lb of salt per 5 gallons of water enters at the rate of 3 gallons per minute. This mixes with what's in the vat, and at the other end, the mixture leaves at the same rate. Find the differential equation that models this situation and its solution, find the amount of salt in the mixture after 1 hour, and find the limiting amount of salt in the vat if the process runs indefinitely.
29. A vat with a capacity of 8000 liters originally contains 1000 liters of water in which 300 g of sugar is mixed. At one end, a mixture of water containing 4 g of sugar per liter of water enters at the rate of 6 liters per minute. This mixes with what's in the vat, and at the other end, the mixture leaves at the rate of 4 liters per minute. Find the differential equation that models this situation and its solution, find the amount of sugar in the mixture after 1 day, and find the amount of sugar in the vat when it reaches full capacity.
30. A mass weighing 2 lbs stretches a spring 1 inch (1/12 feet). The mass is pulled down 4 more inches (1/3 foot) then released. When the mass is moving at 2 feet/second, the surrounding medium applies a resistance force of 2 lbs. Find the initial value problem that governs the motion of the bobbing mass and solve for $u(t)$. Is this system underdamped, critically damped, or overdamped?

Find the Laplace Transforms, $H(s) = L\{f(t)\}$:

31. $L\{4\}$
32. $L\{-2\}$
33. $L\{e^{3t}\}$
34. $L\{5e^{-4t}\}$
35. $L\{9t\}$
36. $L\{2t^2\}$
37. $L\{t^3 - 4t^2 + t - 6\}$
38. $L\{(t+3)(t^3 - 2)\}$
39. $L\{\cos 4t\}$
40. $L\{\sin 8t\}$
41. $L\{e^{2t} \cos 6t\}$
42. $L\{\cos^2 10t\}$
43. $L\{(t-6)^2\}$
44. $L\{u_2(t)(t-2)^4\}$
45. $L\{u_1(t)t^3\}$

Solve the following Laplace Transform inversions, $y = f(t) = L^{-1}\{H(s)\}$:

46. $L^{-1}\left\{\frac{3}{s}\right\}$
47. $L^{-1}\left\{\frac{6}{s^2}\right\}$

48. $L^{-1}\left\{\dfrac{3}{s-5}\right\}$

49. $L^{-1}\left\{\dfrac{10}{s^5}\right\}$

50. $L^{-1}\left\{\dfrac{1}{(s+1)(s-6)}\right\}$

51. $L^{-1}\left\{\dfrac{1}{s^2-s-30}\right\}$

52. $L^{-1}\left\{\dfrac{1}{s^2+25}\right\}$

53. $L^{-1}\left\{\dfrac{s}{s^2+18}\right\}$

54. $L^{-1}\left\{\dfrac{s-3}{(s-3)^2+36}\right\}$

55. $L^{-1}\left\{\dfrac{s}{(s-4)^2+81}\right\}$

56. $L^{-1}\left\{\dfrac{1}{s^3+4s}\right\}$

57. $L^{-1}\left\{\dfrac{s-2}{(s+1)^2+49}\right\}$

58. $L^{-1}\left\{\dfrac{1}{s^2+4s+8}\right\}$

59. $L^{-1}\left\{\dfrac{e^{3s}}{s^2+50}\right\}$

60. $L^{-1}\left\{\dfrac{e^{-s}+e^{-5s}}{s^2+9s+20}\right\}$

Use Laplace Transforms to solve:

61. $y'' + 3y' - 18y = t$, $\quad y(0) = 2, y'(0) = -1$
62. $y'' - y' - 56y = e^{3t}$, $\quad y(0) = 1, y'(0) = 4$
63. $y'' + 16y = \begin{cases} 2, & t < 2 \\ t, & t \geq 2 \end{cases}$, $\quad y(0) = 1, y'(0) = 2$

Solve the following differential equations written as systems:

64. $\mathbf{x}' = \begin{bmatrix} 2 & 4 \\ 1 & 5 \end{bmatrix} \mathbf{x}$

65. $\mathbf{x}' = \begin{bmatrix} 3 & -1 \\ 2 & 0 \end{bmatrix} \mathbf{x}, \quad \mathbf{x}(0) = \begin{bmatrix} 3 \\ 2 \end{bmatrix}$

66. $\mathbf{x}' = \begin{bmatrix} 1 & -1 \\ 1 & 1 \end{bmatrix} \mathbf{x}$

67. $\mathbf{x}' = \begin{bmatrix} 2 & -1 \\ 1 & 4 \end{bmatrix} \mathbf{x}$

68. $\mathbf{x}' = \begin{bmatrix} 1 & 2 & 0 \\ 0 & 3 & 1 \\ 0 & 0 & 2 \end{bmatrix} \mathbf{x}$

Solutions

1. $y = Ce^{10x}$
2. $y = Ce^{-2x^2}$
3. $y = \dfrac{3}{1-3\ln x}$
4. $y = -\dfrac{1}{2} + Ce^{2x^2}$
5. $y = \dfrac{24x-3}{64} + Ce^{-8x}$
6. $y = \dfrac{1}{36}(x^3 + C)^2$
7. $y = \dfrac{1}{6}\left(1 + 23e^{-3x^2}\right)$
8. $y = C_1 e^{5x} + C_2 e^{-4x}$
9. $y = \dfrac{1}{2}(5e^{2x} + e^{-6x})$
10. $y = C_1 \cos 5\sqrt{2}x + C_2 \sin 5\sqrt{2}x$
11. $y = C_1 e^{7x} + C_2 x e^{7x}$
12. $y = C_1 e^{-x} \cos \sqrt{7}x + C_2 e^{-x} \sin \sqrt{7}x$
13. $y = C_1 e^{-4x} + C_2 e^x + C_3 e^{5x}$
14. $y = C_1 e^{2x} + C_2 \cos 2x + C_3 \sin 2x$
15. $y = C_1 e^{5x} + C_2 e^{-x} + C_3 x e^{-x}$
16. $y = C_1 \cos 3x + C_2 \sin 3x + C_3 x \cos 3x + C_4 x \sin 3x$
17. $y = C_1 e^{-2x} + C_2 e^{-x} + \dfrac{1}{42} e^{5x}$
18. $y = C_1 e^{-2x} + C_2 e^{4x} - \dfrac{1}{2} x e^{-2x}$
19. $y = C_1 \cos 4x + C_2 \sin 4x - \dfrac{3}{8} x \cos 4x$
20. $y = C_1 e^{-3x} + C_2 e^{5x} - \dfrac{4}{15}x^2 + \dfrac{1}{225}x - \dfrac{122}{3375}$
21. $y = \dfrac{5}{x(1-5\ln x)}$
22. $y = C_1 x^{-1} + C_2 x^{-1} \ln x$
23. $y_2 = t^{-3}; \quad y = C_1 t^3 + C_2 t^{-3}$
24. $0 < x < \sqrt[3]{e}$, or $0 < x < 1.3956 \ldots$
25. $0 < x < \sqrt[5]{e}$, or $0 < x < 1.2214 \ldots$
26. $P(t) = \dfrac{1}{0.0002 - 0.00000645t}$; $P(11) \approx 7750$ people
27. $V(t) = \left((1.926 \times 10^{20})t + 1.6 \times 10^{21}\right)^{1/4}$; $V(12) \approx \$250{,}079$
28. $Q' + \dfrac{3}{2000}Q = \dfrac{6}{5}$, $Q(0) = 50$; $\quad Q(t) = 800 - 750 e^{(-3/2000)t}$; $Q(60) \approx 114$ lbs; $Q(t \to \infty) = 800$ lbs.
29. $Q' + \dfrac{2}{500+t}Q = 24$, $Q(0) = 300$; $Q(t) = 4000 + 8t - 3700(500)^2(500+t)^{-2}$; $Q(1440) \approx 15{,}274$ g; $Q(3500) \approx 31{,}942$ g.

30. $u'' + 16u' + 384u = 0$, $u(0) = \frac{1}{3}$, $u'(0) = 0$;

$u(t) = \frac{1}{3}e^{-8t}\cos 8\sqrt{5}t + \frac{\sqrt{5}}{15}e^{-8t}\sin 8\sqrt{5}t$; underdamped.

31. $H(s) = \frac{4}{s}$

32. $H(s) = -\frac{2}{s}$

33. $H(s) = \frac{1}{s-3}$

34. $H(s) = \frac{5}{s+4}$

35. $H(s) = \frac{9}{s^2}$

36. $H(s) = \frac{4}{s^3}$

37. $H(s) = \frac{6}{s^4} - \frac{8}{s^3} + \frac{1}{s^2} - \frac{6}{s}$

38. $H(s) = \frac{24}{s^5} - \frac{18}{s^4} - \frac{2}{s^2} - \frac{6}{s}$

39. $H(s) = \frac{s}{s^2+16}$

40. $H(s) = \frac{8}{s^2+64}$

41. $H(s) = \frac{s-2}{(s-2)^2+36}$

42. $H(s) = \frac{1}{2}\left(\frac{s}{s^2+400} + \frac{1}{s}\right)$

43. $H(s) = \frac{2}{s^3} - \frac{12}{s^2} + \frac{36}{s}$

44. $H(s) = \frac{24e^{-2s}}{s^5}$

45. $H(s) = e^{-s}\left(\frac{6}{s^4} - \frac{6}{s^3} + \frac{3}{s^2} - \frac{1}{s}\right)$

46. $y = 3$

47. $y = 6t$

48. $y = 3e^{5t}$

49. $y = \frac{5}{12}t^4$

50. $y = \frac{1}{7}(e^{6t} - e^{-t})$

51. $y = \frac{1}{11}(e^{6t} - e^{-5t})$

52. $y = \frac{1}{5}\sin 5t$

53. $y = \cos 3\sqrt{2}t$

54. $y = e^{3t}\cos 6t$

55. $y = e^{4t}\cos 9t + \frac{4}{9}e^{4t}\sin 9t$

56. $y = \frac{1}{4}(1 - \cos 2t)$

57. $y = e^{-t}\cos 7t - \frac{3}{7}e^{-t}\sin 7t$

58. $y = \frac{1}{2}e^{-2t}\sin 2t$

59. $y = u_3(t)\left(\frac{\sqrt{2}}{10}\sin\left(5\sqrt{2}(t-3)\right)\right)$

60. $y = u_1(t)(e^{4-4t} - e^{5-5t}) + u_5(t)(e^{20-4t} - e^{25-5t})$

61. $y = \frac{100}{81}e^{3t} + \frac{251}{324}e^{-6t} + \frac{1}{18}t + \frac{1}{108}$

62. $y = \frac{56}{75}e^{8t} + \frac{41}{150}e^{-7t} - \frac{1}{50}e^{3t}$

63. $y = \frac{1}{2}\sin 4t + \cos 4t + u_2(t)\left(\frac{1}{16}(t-2) + \frac{31}{64}\sin(4t-8) + \frac{1}{64}\cos(4t-8)\right)$

64. $y = C_1\begin{bmatrix}1\\1\end{bmatrix}e^{6t} + C_2\begin{bmatrix}-4\\1\end{bmatrix}e^t$

65. $y = 4\begin{bmatrix}1\\1\end{bmatrix}e^{2t} - \begin{bmatrix}1\\2\end{bmatrix}e^t$

66. $y = C_1 e^t\begin{bmatrix}-\sin t\\\cos t\end{bmatrix} + C_2 e^t\begin{bmatrix}\cos t\\\sin t\end{bmatrix}$

67. $y = C_1\begin{bmatrix}-1\\1\end{bmatrix}e^{3t} + C_2\left(\begin{bmatrix}-1\\1\end{bmatrix}te^{3t} + \begin{bmatrix}0\\1\end{bmatrix}e^{3t}\right)$

68. $y = C_1\begin{bmatrix}1\\1\\0\end{bmatrix}e^{3t} + C_2\begin{bmatrix}-2\\-1\\1\end{bmatrix}e^{2t} + C_3\begin{bmatrix}1\\0\\0\end{bmatrix}e^t$

Report any errors or suggestions to the author at www.surgent.net/debook.

www.ingramcontent.com/pod-product-compliance
Lightning Source LLC
Chambersburg PA
CBHW071358210526
45465CB00001B/159